# ELEMENTS OF
# ALGEBRA

# ELEMENTS OF
# ALGEBRA

BY

## HOWARD LEVI

PROFESSOR OF MATHEMATICS
SCHOOL OF GENERAL STUDIES
COLUMBIA UNIVERSITY

## FOURTH EDITION

CHELSEA PUBLISHING COMPANY
NEW YORK

# PREFACE TO THE FOURTH EDITION

THE PRESENT EDITION is substantially the same as the third. A number of new exercises have been added, which have kindly been supplied by Mr. Paul Meyer.

# PREFACE TO THE SECOND EDITION

THIS BOOK IS the text for a mathematics course which has been given for the past six years in the School of General Studies of Columbia University. The course is not at all conventional, departing sharply from tradition both in subject matter and approach. Now that the course has been absorbed into the curriculum of the school it is possible to see, in usable perspective, how the novelties of the course acted on its students and on its teachers.

The course has two educational objectives. One is to provide the first of a sequence of mathematics courses leading to advanced work in mathematics and related subjects. The other is to insure that the liberal arts student, whether he takes the course out of curiosity or out of the compulsions of a degree requirement, will be exposed to genuine mathematical problems and procedures. The unconventional character of the course is a consequence of the attempt to attain these objectives, for neither of them is achieved by traditional introductory courses. For several reasons, some relating to the composition of our own student body, and some relating to the fate of mathematics education in the secondary schools throughout the nation, it was also proposed that absolutely no mathematics training beyond simple arithmetic be required. On the other hand, since our students are adults, we can and do require that they be willing to work hard at mastering fairly subtle concepts and fairly elaborate arguments.

Our experiences with the course over the last six years enable us to recognize and sometimes to correct some unprofitable tendencies associated with taking it. For instance, it appears that there will always be some students who consider that the effort to define "2" and "4" and "+" and to prove that "2 + 2 = 4" is not only unnecessary but is, in fact, downright insulting. This attitude seems to go with a generally anti-intellectual position and does not lead to a happy relation between the student and his work or his teacher. There is another extreme position adopted by some students which also leads to mathematical stagnation. The keynote of this position is "Euclid alone has looked on Beauty bare." The student with this attitude shirks the actual details of the development of the algebraic structures and floats on a sea of generalities. While this does not have as unpleasant side effects as the first attitude it does not have satisfactory results. Teachers, too, have been known to fall short of perfection. Some teachers are so impressed by the abstractness of the subject (they learned it themselves in graduate school) that they spend interminable hours "motivating" the student with material far less cogent and intelligible than the subject matter itself. Others find it hard to keep in mind that some of their students really have not studied algebra before, and their lapses plant anxieties in these students which interfere seriously with the learning process. In all these situations the most effective procedure seems to be a resolute insistence on the actual subject matter, rather than specific counter-measures.

Of course there are more substantial aspects to the appraisal of this new course than examining the emotions it evokes from students and teachers. Not only has the course introduced topics not ordinarily taught to beginners but it has discarded topics which have been taught for over a century. While there is abundant evidence of the need for these changes, it must also be recognized that any innovation is likely to produce uneasiness in the mind of the conservative whose important office it is to respect and preserve whatever has made a place for itself. He feels that even if no visible value adheres to a certain part of a process, discarding this part might break some necessary relation and injure the whole process. From this point of view the man who rows up a stream at three miles an hour and down the stream at six miles an hour is suspected of carrying some precious cargo in his boat which makes the enterprise so important that expelling him from the curriculum would be a mistake. We are able to report now that this is not the case. Nothing essential has been lost from the course by eliminating the man, the boat, and other paraphernalia of traditional elementary algebra courses. Some students who began their mathematical careers in the course are now successful graduate students in mathematics. Many liberal arts students are grateful for the enlightening experiences they had while taking it. This is why we consider the course to be successful; because those who take it see and do enough mathematics so that they can form a just notion of what the subject is and what its practitioners are concerned with, and so that those who want to continue have taken a significant step along the way.

New York, 1956.

**H. L.**

# PREFACE TO THE FIRST EDITION

SEVERAL STIPULATIONS guided the selection of material for this book. The first was, that nothing should be taught in a mathematics class that a qualified instructor cannot believe and become at least mildly excited about. This stipulation excludes a great deal of material that is traditionally offered in college mathematics courses, for it is a curious historical accident that many mathematics departments are committed to teaching subjects that have no useful relation to any valid mathematical enterprise.

Another stipulation was that the material should not be diluted or otherwise debased in an effort to make it alluring or give it a spurious note of accessibility. Whimsy, and similar bids for the student's indulgence, were to be avoided. It was to be understood that mathematics as we know it today is the accumulated and distilled product of many profound minds working for many years, and that the student's exposure to this concentrated product might tend to upset him. Since the student's experience, judgment, and taste had no influence on the development of mathematics, he should not expect that the subject will tell him what he already knows, or what he thinks he should know, or even what he would like to know. The student should be required to bring a respectful curiosity to the subject. He should in turn expect that reasonable effort on his part will bring him in contact with some of the intellectual triumphs which compose mathematics. He should not expect to improvise mathematics, nor should he be put off in his study of mathematics with puzzles and busy work.

I want to thank the following present and former instructors in the School of General Studies for suggestions arising from their successful efforts to establish the course in that school: Drs. W. Wernick, R. H. Brown, G. Raney, A. Hillman, and F. Steinhardt; Mrs. R. Strodt; Mr. C. W. Langley. I want to thank Mrs. Strodt for making available some exercises she devised while teaching the course.

New York, 1954

**H. L.**

# PREFACE TO THE PRELIMINARY EDITION

This book is the text for a one-semester three-point mathematics course given in the School of General Studies of Columbia University. The description of the course given in the Announcement of the School of General Studies is:

This course carries out the construction of the natural numbers, the integers, the rationals, and the reals, and develops the algebra appropriate to each of these number systems. The course undertakes to provide a firm and coherent basis for further study of mathematics, and, incidentally, to give the student whose chief interests lie elsewhere, an acquaintance with genuine mathematical problems and procedures.

No previous training in algebra is presupposed. The only prerequisite is some curiosity about the nature of mathematics.

In the past three years, over 2000 students have taken the course, and although one student attributed a heart attack to the experience, most of the students ultimately made sense of the material. It is true that the material is quite different from what is taught in high schools under the name of algebra. The student is urged not to use this difference as an excuse for sulking; he is invited instead to rejoice that he can confront, without arduous preliminaries, some of the richest and boldest intellectual achievements of the human race.

New York, 1953

**H. L.**

# CONTENTS

# ELEMENTS OF
# ALGEBRA

# CHAPTER I

## SETS, STATEMENTS, AND VARIABLES

### Introduction

ALGEBRA, LIKE ALL OTHER SCIENCES, is a collection of statements. It differs from other sciences in the kind of statements it makes and in the kind of object the statements are about. It is, in fact, a new language. This chapter prepares the reader to understand the special statements and subject matter that make up this new language. It will be discovered that this new language strives at all costs for extreme clarity in its meanings and for complete certainty in its conclusions. Since most of the circumstances which beset us in our daily lives are neither clear nor certain, we must not expect everyday experience to figure in our discussions. The field of vision of algebra is very small, but its focus is wonderfully sharp.

We use the new language to describe the objects we study (mostly numbers) and to formulate and derive their properties. The role of our everyday language will be to formulate our intentions and goals and to appraise our achievements.

### Sets and their Members

In algebra we use the concepts of logic and some of the terms used to convey those concepts, such as *and, or, if, then, all, some, is,* and *not.* We use two other terms of everyday language, *set* and *member of a set,* with slightly changed meanings. In the first place, we shall not require the members of a set to be alike or to be matched in any way. Thus, a goldfish, a roller skate, and an automobile could together constitute a perfectly acceptable set. What is more important is that we shall insist on knowing with complete certainty exactly what things are members of our sets. Thus we will not discuss such sets as "the set of loyal citizens" or "the set of beautiful paintings" because we cannot be certain about the membership lists of these sets. Indeed, if the reader will reflect about any set at all that occurs to him, he will see that it probably does not qualify as the kind of set we propose to discuss, because its membership list is too vague.

While we do not define the notion of *set* in general, we do define individual sets; to *define a set* will mean to specify those things which are its members. It is conceivable that a given set could have several different definitions of this sort. We understand that if two different definitions single out the same members, then the sets so defined are to be considered as the same set.

One set about whose membership list there can be no quibbling is the set with no members at all. This set is called the *empty set,* and we shall refer to it often. It is doubly useful. Its membership list is definite enough to meet our standards; in addition, it is an object that has been specified clearly, and we can use it as a member of other sets.

If $A$ and $B$ are sets and if every member of $A$ is also a member of $B$, we shall say that $A$ is a *subset* of $B$. The notion of a subset of a set is, roughly, the counterpart for sets of the everyday concepts of part and whole. This similarity is marred, however, by the fact that our definition of subset implies that every set is a subset of itself. The notion of proper subset gives us a closer analogy to the part-whole concept: We say that a set $A$ is a *proper subset* of a set $B$ if $A$ is a subset of $B$ but $B$ is not a subset of $A$.

We are now able to make some statements with completely certainty. One is, that *if $A$ is a set, then the empty set is a subset of $A$.* Another is, that *if $A$ is a set, then $A$ is a subset of $A$.* Another is, that *if $A$ is a set, then $A$ is not a proper subset of $A$.* Each of these statements is a necessary consequence of our definitions.

The reader may be willing to concede that certainty is attainable in discussing sets and their members but may feel that it is achieved by attending to utterly trivial matters. To show him that fairly subtle concepts can be expressed in terms of sets and their members, we consider the problem of defining, for someone who cannot count, the idea of a set with one member. We suggest this:

$A$ is a set with one member if

    a) $A$ is a set,
    b) $A$ is not the empty set,
    c) if $B$ is a proper subset of $A$, then $B$ is the empty set.

## Construction of Sets

Since most familiar sets are not specified clearly enough for our purposes, we have to construct our own sets. We now discuss some schemes for constructing sets.

If we are given some sets, we shall mean by their *union* the set whose members are those things that are members of at least one of the given sets. We shall mean by the *intersection* of the given sets, the set whose members are those things that are members of all the given sets. We use special symbols to designate unions and intersections. The union of set $A$ and set $B$ is designated by $A \cup B$ (read: $A$ union $B$), and their intersection by $A \cap B$ (read: $A$ intersection $B$).

*Example*: Let $A$ be the set whose members are the numbers 1, 2, 3, 4, and let $B$ be the set whose members are the numbers 3, 4, 5. Then $A \cup B$ is the set whose members are 1, 2, 3, 4, 5; and $A \cap B$ is the set whose members are 3 and 4.

A somewhat more elaborate scheme for constructing sets is based on the concept of an ordered pair. By an *ordered pair* we mean the combination of

    (a) a set formed by selecting some thing and then again selecting some thing (not necessarily a different thing)
    (b) a specification of one of the selected things as the "first."

An ordered pair differs from an ordinary pair in that its members need not be different and also in that one of its members is singled out as the first. An example of an ordered pair is the pair the planet Mars and the planet Venus, with Mars as first. Another example is the planet Jupiter, the planet Jupiter again (with the planet Jupiter necessarily first). It is customary to write ordered pairs with parentheses and a comma, our first example being written (Mars, Venus), our second (Jupiter, Jupiter). Throughout this book a symbol such as $(a, b)$ will mean the ordered pair consisting of the thing $a$, the thing $b$, with $a$ designated as first.

We have already encountered two ways of constructing new sets from given sets: forming their union and forming their intersection. We now introduce a third way of constructing a new set from two given sets.

The *product set* of set $A$ and set $B$ is the set of all ordered pairs $(a, b)$, where $a$ is a member of set $A$ and $b$ is a member of set $B$. The product set is denoted by the special symbol $A \times B$ (read: $A$ cross $B$).

*Example*: Let $A$ be the set whose members are the numbers 5 and 7, and let $B$ be the set whose members are the numbers 3, 4, and 5. Then $A \times B$ is the set whose members are the ordered pairs $(5, 3)$, $(5, 4)$, $(5, 5)$, $(7, 3)$, $(7, 4)$, $(7, 5)$.

Note that while every product set is a set whose members are ordered pairs, not every set of ordered pairs is a product set. For instance, the set of ordered pairs $(5, 3)$, $(7, 3)$, $(7, 4)$, $(7, 5)$ is not a product set. Any product set which contains $(5, 3)$ and $(7, 4)$ must also contain $(5, 4)$, and this set does not.

Still another way of forming new sets from given sets is the device of using the given sets themselves as members of a set. Thus, if $A$ and $B$ are sets, the set whose members are $A$ and $B$ is a new set. We shall use this device later with a single set and ask the reader to distinguish between (i) a set $A$, and (ii) the set whose only member is that set $A$. Note that the former may have many members, whereas the latter has only one.

It is customary to use braces { } to indicate members of a set. Thus, $\{a, b, c\}$ means the set whose members are $a$ and $b$ and $c$, and

$\{(1, 2), (1, 3)\}$ means the set whose members are the ordered pairs $(1, 2)$ and $(1, 3)$. If $A$ is a set, $\{A\}$ means the set whose only member is the set $A$.

## Variables and Statement-forms

Almost everyone knows that algebra has something to do with $x$, the unknown quantity. We are now going to describe a way of making statements which uses symbols like $x$ in precisely the way they are actually used in algebra.

Many statements assert something about each of the members of a given set. For instance, the statement "All men are mortal" can be regarded as meaning

> "Adam is mortal, and
> Cain is mortal, and
>
> . . . . . . . . . . . . . . . . . . .
>
> Socrates is mortal, and
>
> . . . . . . . . . . . . . . . . . . .
>
> John Jones is mortal, and
> etc."

That is, the statement "All men are mortal" is a concise way of stating simultaneously each of a long list of statements about individual men. The statements in question are all those obtainable by filling in the name of a man in the blank space of the form "——— is mortal." It should be noted that the expression "——— is mortal" is not a statement. It becomes a statement when we put a name in the blank space. We shall call such an expression a *statement-form*.

It is customary to write some symbol such as $x$ in the blank space of a statement-form. Thus, "——— is mortal" becomes "$x$ is mortal." Again we have only a statement-form, not a statement; the symbol $x$ plays exactly the same role as the blank space. Such a symbol is called a *variable*. However, the statement "All dogs are mortal" may also be considered as standing for a long list of statements, "Rover is mortal, and Fido is mortal, and etc.," all likewise of the form "$x$ is mortal." We see that while the original statement leads to a definite statement-form, the same statement-form can serve for many different statements. It follows that a statement-form cannot, by itself, adequately convey the meaning of a statement. The difficulty is that we cannot tell what the variable in a statement-form is supposed to designate, merely by looking at the statement-form. The solution of this difficulty is quite simple. We agree, when we use a statement-form with a variable, to describe in a separate phrase the set to which the variable refers. This set is known as the *domain* of the variable. Our final restatement of "All men are mortal" is "$x$ is mortal, where $x$ is a variable whose domain is the set of men."

A variable may occur more than once in a statement-form. An instance of this is furnished by "If $x$ is even, then $x + 1$ is odd." We require, in such a case, that if the variable is to be replaced by a member of its domain, then the same replacement must be made throughout the statement-form. That is, if a variable occurs several times in a statement-form, it should not be replaced by one thing in one part of the statement-form and by a different thing in another.

These comments are intended to show how we propose to use variables in algebra. We do not suggest that the illustrative statement "All men are mortal" is a particularly clear one, not having an acceptable definition of "men" or of "mortality." We do contend that if the domain of the variable is an acceptable set, then this procedure produces acceptable statements. The reader may be unable to see that anything is gained by this kind of restatement. The point of using the statement-form procedure is that in algebra, statements put in this way can be processed in a standardized way to obtain further statements. Indeed, the whole body of statements which constitute algebra can be deduced from a very few statements made in this way.

A statement-form may contain several variables. An example of such a statement-form is "$x$ is greater than $y$." For this to acquire a meaning, a domain would have to be specified for each variable.

For instance, $x$ might have as domain the set whose members are 7 and 8, and $y$ might have as domain the set whose members are 2, 3, and 4. With these specifications we have a statement that means

> "7 is greater than 2, and
> 7 is greater than 3, and
> 7 is greater than 4, and
> 8 is greater than 2, and
> 8 is greater than 3, and
> 8 is greater than 4."

It should be noted that for each variable we substitute a member of its domain in all possible ways, and that all of the statements so obtained are asserted. If one or more of the statements obtained by this substitution process is false, we say that the original statement is false. If all are true, we say that the original statement is true.

Most algebraic equations are statement-forms. The separate phrase that should accompany the variables and state to what they refer is rarely given. The fact is that the variables refer almost always to numbers. Mention of this may be omitted if it is agreed once and for all that this is the case, but all too often mention of it is absent, not by agreement, but from ignorance. The student then concludes that the letters in his equations have some mysterious life of their own, whereas in fact it is the properties of the numbers designated by these letters which justify the proceedings of algebra.

## Functions

For the last two centuries, the concept of *function* has been central in mathematics. We are going to describe in everyday language the special meaning this term has in mathematics. We will then give a formal definition of function entirely in our set language. This will not only prepare the reader for our later work but will also provide him with another instance of the interplay of the two languages in the development of algebra.

Generally, we say we have a *function* when two sets $A$ and $B$ are given and a member of $B$ is assigned to each member of $A$. For instance, $A$ could be a set of people, $B$ could be the alphabet, and to each person of $A$ might be assigned the first letter of his last name. Or, $A$ could be a set of married men, $B$ the set of all women, and his wife might be assigned to each member of $A$. Note that $A$ and $B$ are not subject to the same requirements. To each member of $A$ there must be assigned one member of $B$, called its *image*. Every member of $A$ must have exactly one image. Not every member of $B$ need be an image, and a member of $B$ can be the image of several members of $A$. As for the scheme of making the assignments, it can have some underlying regularity, as in the above examples, or it can be altogether arbitrary, as in the example below. The only condition is that the assignments be made.

We now give our formal definition.

DEFINITION: Let $A$ and $B$ be sets. By a *function on $A$ to $B$* we mean any subset of $A \times B$ in whose ordered pairs each member of $A$ appears exactly once as a first member.

*Example*: Let $A$ be the set whose members are 5 and 7, and $B$ be the set whose members are 3, 4, and 5. Then the subset of $A \times B$ whose members are $(5,3)$, $(7,4)$ is a function on $A$ to $B$. Another example of a function on $A$ to $B$ is the subset of $A \times B$ whose members are $(5,4)$ and $(7,4)$. There are in all nine functions on $A$ to $B$, namely the following:

$$(5,3) \text{ and } (7,3),$$
$$(5,3) \text{ and } (7,4),$$
$$(5,3) \text{ and } (7,5),$$
$$(5,4) \text{ and } (7,3),$$
$$(5,4) \text{ and } (7,4),$$
$$(5,4) \text{ and } (7,5),$$
$$(5,5) \text{ and } (7,3),$$
$$(5,5) \text{ and } (7,4),$$
$$(5,5) \text{ and } (7,5).$$

Functions are frequently designated by letters of the alphabet, and it is customary to use the letter that designates a function to help indicate the assignments it makes: If $A$ and $B$ are sets and if $f$ is a function on $A$ to $B$, then for each $a$ of $A$, $f(a)$ means that member of $B$ which is assigned to $a$ by the function $f$. If, for instance, the last function listed above is designated by $f$, then $f(5)$ is 5 and $f(7)$ is 5. If the next to the last is designated by $g$, then $g(5)$ is 5 and $g(7)$ is 4.

Let $A$ and $B$ be any two sets. It may be possible to pair the members of $A$ with the members of $B$ in such a way that

(1) every member of $A$ is paired with a member of $B$;
(2) every member of $B$ is paired with a member of $A$;
(3) different members of one set are not paired with the same member of the other set.

Such a pairing off of the members of two sets is called a *one-to-one correspondence* between the two sets.

*Example*: Let $A$ be the set {#, \$, *}, and let $B$ be the set {$x, y, z$}. We can establish a one-to-one correspondence between $A$ and $B$ in many ways, among them the following:

$$\text{\$ paired with } x,$$
$$\text{\# paired with } y,$$
$$\text{* paired with } z.$$

Associated with each one-to-one correspondence between $A$ and $B$ is a special and important function on $A$ to $B$. For if we construct ordered pairs by coupling each member of $A$ with its correspondent in $B$, we obtain a subset of $A \times B$ that meets the definition of function on $A$ to $B$. However, not all functions on $A$ to $B$ arise from one-to-one correspondences. In general, a function on $A$ to $B$ pairs each member of $A$ with some member of $B$. This pairing can fail to be a one-to-one correspondence in one of two ways: the same member of $B$ might be paired with more than one member of $A$, or some members of $B$ might be paired with no member of $A$.

It is very unlikely that a beginner in mathematics could see how this idea of function could be the principal concept in the development of a science. Most of the remainder of this book is built up around this idea and we hope that the reader's curiosity is great enough for him to continue till he sees for himself how central this idea is. Each of our analyses of addition and multiplication, as these operations occur in the various number systems we study, is based on the concept of function.

## Rules of Inference and Proofs

So far we have analyzed statements from what may be called a grammatical point of view; we have inquired into ways of conveying their meaning. A much more important problem is raised when we inquire into the truth of statements. We can all agree that the statements "Nobody can swim the Atlantic Ocean" and "No circle is a square" are both true. It is even possible to believe in the truth of these statements with equal intensity, but the ways of deciding on the truth of the statements are surely quite different. Had the Atlantic been narrower or the human race tougher the first might have been false. It is a matter of fact and is tested by experiment. The second statement has a different kind of justification, based on logical proof, and denying it does violence to the most fundamental demands of logic and common sense. Most of the statements of mathematics are of this latter sort. They are supposed to have a justification separate from the fact that we may believe them to be true or that we have tested them experimentally. For instance when we discuss the statement "$2 + 2 = 4$" later on we shall see that whether or not we believe the statement is true or whether or not we have tested it experimentally with groups of apples has nothing to do with our mathematical handling of it. Instead our task will be to furnish mathematically acceptable proofs of the statements we assert. We consider now what such a proof is supposed to be.

In the first place a mathematical proof is always a proof of a statement. A *step* in a proof consists in adjoining a single statement to a collection of statements already accepted, in accordance with carefully prescribed rules called the *rules of inference*. To prove any given statement means to show that these rules permit the statement in question to be so adjoined. Three of our rules of inference are purely logical rules. One is known variously as *modus ponens* and as the *Law of Detachment*. In order to describe it we refer to certain unspecified statements, which we indicate by "*p*" and "*q*," and to the compound statement "If *p* then *q*." The rule says that the statement "*q*" may be added to the collection of accepted statements if both the statements "If *p* then *q*" and "*p*" are already in the collection. For instance, if the two statements "If 4 is even then 5 is odd" and "4 is even" are in the collection of accepted statements, then the statement "5 is odd" can be added to this collection. Another rule of inference is called *proof by contradiction*. If a statement "*p*" has the property that its addition to the collection of accepted statements makes it possible both to prove some statement "*q*" and to prove its denial "not *q*," then we say that the statement "*p*" *leads to a contradiction*. The law of contradiction permits the adjoining to the collection of accepted statements the denial "not *p*" of any statement "*p*" which leads to a contradiction. For instance we prove later that the number 2 does not have a certain property by showing that if we assume that it does have this property then it can be proved of a certain other number both that it is even and that it is not even. Our remaining logical rule is that any definition can be admitted to the collection of accepted statements.

We also make use of a sweeping rule of inference involving sets. The rule is that any statement about sets and their members which appeals strongly to our sense of what is correct can be admitted to the collection of accepted statements. For instance, in the next chapter we accept without argument certain statements about finite sets even though these are quite unfamiliar, and in a context, moreover, in which we are proving what appears to be quite familiar facts about numbers. This somewhat casual attitude—called the *naive,* or *intuitive,* use of sets—has a respectable place in mathematical investigations, even though mathematicians have shown that it can lead to paradoxes. There do exist more refined ways of using sets but their study is best deferred until the student has some mathematical maturity. In the meantime he can be confident that this sweeping rule about the use of sets will be invoked in moderation and will lead only to civilized portions of mathematics.

It often happens that a statement whose proof is sought cannot be added to the existing collection of accepted statements by a single application of a rule of inference. The usual procedure then is to try to enlarge the body of accepted statements so as to permit the addition of the given statement. This requires a search for suitable auxiliary statements to add to the original collection. Such a procedure gives our proofs a uniform structure. They all consist of repeated use of a single step: that of adjoining a single new statement to the body of statements already accepted. To understand a proof requires the recognition that each step in the sequence is a legitimate instance of an application of a rule of inference.

To devise an original proof of a given statement requires more than this passive recognition. There are usually many different statements which are immediate consequences of the pool of currently accepted statements, and it is not always easy to pick the relevant ones which lead to the statement to be proved. This aspect of a proof is not at all mechanical; this is the area in which mathematical talent makes itself known, and where experience is a prerequisite to facility. Perhaps the hardest of all proof activities is the art of conjecturing what statements are provable. The reader of this book is chiefly required to understand proofs already given. Very occasionally he is asked to devise his own proof of a statement. Perhaps half a dozen times in the- whole course he is invited to conjecture a timely and provable statement.

Important proved statements are called *theorems.* Proved statements which are preliminary to more important theorems are sometimes called *lemmas.* Proved statements which follow easily from, or are special cases of theorems are sometimes called *corollaries.*

## Beliefs, Validity, and Sets

It will help the beginner to understand mathematics if he tries to acquire a suitable attitude toward carefully expressed definitions and logically sound proofs. In this connection it is useful to consider how what he believes to be familiar and true is related to the body of definitions and statements he encounters in mathematics. In the first place he will discover that his belief in the truth of a mathematical statement cannot be offered as a substitute for a proof of the statement, nor does the fact that a statement has been proved mean that it is true and he therefore must believe it. A statement which has been proved is called *valid,* and there is no simple connection between the truth of a statement and its validity. This is not to suggest that there is no place in mathematics for beliefs, but rather that the place where belief is relevant has been shifted from the statements themselves to the connections between statements. Anyone who studies a proof has to decide whether he believes that each step is a legitimate application of a rule of inference. He does not have to decide whether he believes that the statements which are involved in the steps are true.

It can be readily appreciated that decisions about validity are easier to make if the statements in question are expressed clearly and in a neutral language rather than in one which tends to bias judgment by invoking hidden associations and prejudices. This can help to account for the popularity among mathematicians of the use of sets. It turns out that almost all mathematics can be formulated in terms of sets. Most mathematical definitions can be regarded as definitions of sets and most mathematical statements can be regarded as assertions that a certain object is a member of a certain set. By voluntarily impoverishing his language to this level a mathematician protects himself from the intrusion of extraneous ideas into his deductive system. This is not to suggest that discoveries in mathematics must take place in such a circumscribed framework. The most extravagant and poetic flights of fancy can contribute to the invention of new mathematics. All that is required is that, whatever may have been the episodes in its discovery, each statement must submit successfully to this set-theoretic treatment before it can be considered to be part of the world of mathematics.

## EXERCISES

1. *Sets and their members.*
   a) $A$ is a set whose only member is $a$. List its two subsets.
   b) $A$ is a set whose members are $a$ and $b$. List its four subsets.
   c) $A$ is a set whose members are $a$, $b$, and $c$. List its eight subsets.
   d) The empty set has one subset. Specify that subset and determine if it is a proper subset.

2. *The construction of sets.*
   a) Give an example of a familiar set which is the union of two familiar sets. Do the same for intersection.
   b) If $A$ is a set with seven members and $B$ is a set with five members, what can you say about the number of members of $A \cup B$, $A \cap B$, $A \times B$?
   c) Using only the empty set and sets constructable from it, construct a set with five members.

3. *Variables and statement-forms.*
   a) What is the meaning of
      i) $x$ is an even number, where $x$ is a variable whose domain is the set $\{2, 4, 6, 8\}$.
      ii) $x$ is assigned to $2 + x$, where $x$ is a variable whose domain is the set $\{3, 4, 5, 6\}$.
   b) Express in statement-form variable language:
      i) 3 is greater than 2, and 4 is greater than 2.
      ii) 3 is greater than 2, and 4 is greater than 2, and 3 is greater than 3, and 4 is greater than 3.
      iii) 3 is greater than 2, and 4 is greater than 3, and 5 is greater than 4.

4. *Function.*

   a) Each of the following phrases suggests at least one function. You are asked to discover one and to describe it in both the informal and formal way: State capitals, common nicknames, the average length of life of different kinds of animals, the melting point of metals, addition of whole numbers, multiplication of whole numbers.

   b) The statement "$x$ is assigned to $2 + x$, where $x$ is a variable whose domain is the set $\{3, 4, 5, 6\}$" describes a function. Give a formal description of this function.

   c) Find all six one-to-one correspondences between the set whose members are $p, q, r$ and the set whose members are $P, Q, R$.

   d) Give an example of two familiar sets which can surely be put into one-to-one correspondence with each other and also an example of two familiar sets which surely cannot.

5. *Rules of inference and proofs.*

   $A$, $B$, $C$ are sets.

   a) Prove that $A$ is a subset of $A \cup B$.

   b) Prove that $A \cap B$ is a subset of $A$.

   c) Prove if $A$ is not a proper subset of $A \cup B$, then $B$ is a subset of $A$.

   d) Prove that if $A \cap B$ is not a proper subset of $A$, then $A$ is a subset of $B$.

   e) Prove that if $A$ is a subset of $B$ and $B$ is a subset of $A$, then $A$ and $B$ are the same.

   f) Let $D$ be $A \cup B$ and $E$ be $B \cup C$. Prove that $D \cup C$ and and $A \cup E$ are the same.

   g) Let $D$ be $A \cap B$ and $E$ be $B \cap C$. Prove that $D \cap C$ and $A \cap E$ are the same.

6. Let $A$ be the set $\{p, q, r, s\}$,
   let $B$ be the set $\{s, t, u\}$, and
   let $C$ be the set $\{q, r\}$.

a) Find each of the following sets (by specifying its members).

$C \cup C, C \cap C, C \times C, B \times C, C \times B,$
$(A \cup B) \cup C, A \cup (B \cup C),$
$(A \cap B) \cap C, A \cap (B \cap C),$
$C \cup (B \cap A), (C \cup B) \cap (C \cup A),$
$C \cap (B \cup A), (C \cap B) \cup (C \cap A),$
$C \times (B \cup A), (C \times B) \cup (C \times A).$

b) Find all subsets of $B \times C$ which are functions on $B$ to $C$.

c) Find all subsets of $C \times B$ which are functions on $C$ to $B$.

# CHAPTER II

## CARDINAL NUMBERS

### Introduction

IN THIS CHAPTER, WE make our first efforts at formal construction
of mathematical objects. We are going to define, as carefully as
we can, objects that we shall call "zero," "one," "two," "three," etc.,
and operations that we shall call "addition" and "multiplication."
We are then going to try to prove such familiar-sounding statements as "Two plus two equals four." The student who already
knows that two plus two equals four should not feel that on this
account what we do offers him nothing new. Our "two" and his
are almost certainly different, and so are our "plus" and "equals"
as well as our notion of "knowing" a fact. On the other hand, it
would be grossly inaccurate to claim that there is no connection
whatsoever between the two statements. A great deal of mathematics can in fact be regarded as a logically precise realization of
an experience, a conviction, or a hope.

We are also going to try to prove some facts about the system
of mathematical objects as a whole. This will lead us from arithmetic, where the concern is with individual properties of individual
numbers, to algebra, where the concern is with the structure of
the whole system.

### Standard Sets and Cardinal Numbers

Our work with cardinal numbers is carried out in terms of
certain carefully defined sets called "standard sets." In order to

define these sets we need the notion of the *immediate successor* of a set $A$, which we define to be the set $A \cup \{A\}$. The members of the immediate successor of a set $A$ are thus all the members of $A$ and also $A$ itself, considered as a single thing. The immediate successor of $A$ contains $A$ both as a proper subset and as a member.

We now define our standard sets. The empty set is to be a standard set. So is its immediate successor. So is the immediate successor of this set. In fact, all sets that can be obtained from the empty set by repeated construction of immediate successors are to be standard sets. The sets so constructed, together with the empty set, constitute all the standard sets.

Let us denote the empty set by 0. Then the immediate successor of the empty set is $0 \cup \{0\}$. This latter set, being the union of the set $\{0\}$ with the empty set, is the same as the set $\{0\}$. Let us call this standard set, the set 1. Then the immediate successor of 1 is the set $1 \cup \{1\}$, which is $\{0\} \cup \{1\}$, or $\{0, 1\}$. Let us call this standard set 2. Then the immediate successor of 2 is $2 \cup \{2\}$, which is $\{0, 1\} \cup \{2\}$, or $\{0, 1, 2\}$. Let us call this standard set 3. The immediate successor of 3 is $3 \cup \{3\}$, which is $\{0, 1, 2,\} \cup \{3\}$, or $\{0, 1, 2, 3\}$. Proceeding in this way, we construct the endless list of standard sets, naming them with the familiar names of the numbers of everyday life. We have set ourselves a challenge. We are to prove that our standard sets have properties recognizably like the numbers whose names they share.

We define a *finite set* as one that can be put in one-to-one correspondence with a standard set. We define the *cardinal number* of a finite set to be the name of the standard set with which the one-to-one correspondence can be established.

## Addition of Cardinal Numbers

We are going to define addition of cardinal numbers. Our definition is based on operations with finite sets. Let $a$ and $b$ be any cardinal numbers. Let $A$ and $B$ be finite sets having the cardinal numbers $a$ and $b$ respectively, and having no members in common. The cardinal number of the union of $A$ and $B$ is defined to be the *sum* of the cardinal numbers $a$ and $b$ and is denoted by $a + b$

(read: a plus b). The operation that determines $a + b$ is called *addition*. For any cardinal numbers $a$ and $b$, then, we have defined a cardinal number $a + b$. This cardinal number depends upon $a$ and $b$. It does not depend upon the particular sets $A$ and $B$ used in the construction. It is customary to refer to the cardinal numbers $a$ and $b$ as the *terms* of the sum $a + b$.

Some remarks about notation are called for here, which arise from the fact that an expression like $7 + 5$ is often regarded as a problem in arithmetic, calling for some kind of action. Our particular use of the symbol requires us to think of the expression $7 + 5$ not as a problem but as an expression designating a cardinal number. All the cardinal numbers occur in the list 0, 1, 2, ... , and therefore the cardinal number designated by $7 + 5$ must occur in this list, although the particular arrangement of symbols $7 + 5$ does not occur. It is sometimes desirable to determine which particular member of this list is designated by an expression like $7 + 5$, and it is this determination which is the arithmetical problem. We shall say that a symbol that designates a cardinal number is in *standard form* if it occurs in the list of symbols 0, 1, 2, . . . . Thus, 12 is in standard form while $7 + 5$ is not, although it turns out that 12 and $7 + 5$ designate the same cardinal number. We convey this information by the statement $7 + 5 = 12$. The equals sign ($=$) will appear often in our work. Wherever it connects numbers, it is to mean that the number designated by the symbols on the left is the same number as the number designated by the symbols on the right. In general, any statement of the form $A = B$ is to mean that $A$ and $B$ are names of the same thing. We show later how the standard form of a cardinal number can be determined. What we emphasize here is that there is nothing the matter with a cardinal number that is designated by a symbol not in standard form, and that such symbols designate cardinal numbers as adequately as those in standard form.

## The Commutative and Associative Laws of Addition

We formulate and prove two basic facts about our system of numbers. These facts are taken for granted in the everyday number system.

Theorem I  (The Commutative Law of Addition) :

$$x + y = y + x,$$

*where x and y are variables whose domains are the set of cardinal numbers.*

*Note*: The meaning of this statement is the endless list of statements

$$0 + 0 \; = \; 0 + 0$$

and

$$0 + 1 \; = \; 1 + 0$$

and

$$0 + 2 \; = \; 2 + 0$$

and

. . . . . . . . . . . . . . . .

and

$$55 + 97 = 97 + 55$$

and

. . . . . . . . . . . . . . . .

In undertaking to prove this statement, we are committed to proving each of the infinitely many individual statements. Obviously, we cannot accomplish this by taking each individual statement in turn. In order to show that each of these individual statements is true, we show what is logically equivalent, that no one of them is false. To accomplish this it suffices to show that the assumption that at least one of them is false leads to a contradiction.

*Proof*: If the statement were false, there would be cardinals $a$ and $b$ for which $a + b$ and $b + a$ were different. Let $A$ and $B$ be sets having cardinal numbers $a$ and $b$ respectively, and having no members in common. Then, by definition, $a + b$ is the cardinal number of $A \cup B$, and $b + a$ is the cardinal number of $B \cup A$. But, by definition of union, $A \cup B$ and $B \cup A$ are the same set. Thus, $a + b$ and $b + a$ are cardinal numbers of the same set and are therefore the same. This contradicts the assumption that they are different. Thus the assumption that there are cardinals $a$ and $b$

for which $a + b$ and $b + a$ are different is untenable. The theorem is therefore true.

This theorem is known as the *Commutative Law of Addition*. We shall refer to it as CA.

Our discussion of addition, so far, has been limited to the case of two cardinal numbers. This may seem like a meager accomplishment to the student who is prepared to add a long column of numbers. However, if we examine the familiar process of adding more than two numbers, we observe that it is actually a repetition of the operation of adding exactly two of them. For instance, the sum of 3, 5, and 6 might be found by adding 3 and 5 to obtain 8, and then adding 8 and 6 to obtain 14. A common check is then to add in reverse order—6 and 5 to obtain 11, and then 11 and 3 to obtain 14. Actually, the noteworthy aspect of this example is not so much the device of dealing with three numbers, as the check itself. The statement that the same sum should be obtained by adding in the different orders is a consequence of a basic property of our system of cardinal numbers. In order to express this property, we need a way of describing a sum of cardinal numbers one of whose terms is itself a sum. Let $a$, $b$, and $c$ designate cardinals. We mean by $(a + b) + c$ the cardinal that is the sum whose terms are the cardinal $a + b$ and the cardinal $c$. We mean by $a + (b + c)$ the cardinal that is the sum whose terms are the cardinal $a$ and the cardinal $b + c$. Our statement takes the form:

THEOREM II (The Associative Law for Addition):

$$(x + y) + z = x + (y + z),$$

*where $x$, $y$, $z$ are variables whose domains are the set of cardinal numbers.*

*Note*: The meaning of this statement is the endless list of statements

$$0 + (0 + 0) = (0 + 0) + 0 \qquad \text{and}$$

$$\cdots\cdots\cdots\cdots\cdots\cdots\cdots\cdots$$

$$2 + (5 + 11) = (2 + 5) + 11$$

$$\cdots\cdots\cdots\cdots\cdots\cdots\cdots\cdots$$

*Proof*: As in the proof of Theorem I, we need only show that there is no trio of cardinals $a$, $b$, $c$ for which the corresponding individual statement is false. Let $A$, $B$, and $C$ be sets which have cardinal numbers $a$, $b$, and $c$ respectively, and which are such that no member of one of the sets is a member of another of them. Then

$a + b$      is the cardinal number of $A \cup B$;

$(a + b) + c$ is the cardinal number of the union of $A \cup B$ and $C$;

$b + c$      is the cardinal number of $B \cup C$;

$a + (b + c)$ is the cardinal number of the union of $A$ and $B \cup C$.

If we can show that the union of $A \cup B$ and $C$ is the same set as the union of $A$ and $B \cup C$, it will follow that $(a + b) + c$ and $a + (b + c)$ are cardinal numbers of the same set and hence are the same cardinal number. But it follows from the definition of the union of sets that these two sets are the same. This concludes our proof.

## Multiplication of Cardinal Numbers

We are going to define multiplication of cardinal numbers. Let $a$ and $b$ be any cardinal numbers, and let $A$ and $B$ be finite sets having the cardinal numbers $a$ and $b$ respectively. The *product* of the cardinal numbers $a$ and $b$ is defined to be the cardinal number of $A \times B$ and is denoted by $a \cdot b$ (read: $a$ times $b$). The operation which determines $a \cdot b$ is called *multiplication*. For any cardinal numbers $a$ and $b$, then, we have defined a cardinal number $a \cdot b$. This cardinal number depends upon $a$ and $b$ but does not depend on the particular sets $A$ and $B$ used in the construction. It is customary to refer to the cardinal numbers $a$ and $b$ as the *factors* of $a \cdot b$. The dot which denotes multiplication is almost always omitted when at least one of the factors is designated by a letter. Thus, $a \cdot b$ is written as $ab$, $5 \cdot a$ as $5a$ and $b \cdot 6$ as $b6$. However, for obvious reasons, $5 \cdot 6$ is never written as $56$.

## The Commutative and Associative Laws of Multiplication

The two theorems that follow are the analogues for multiplication of the laws CA and AA, which have already been proved.

Theorem III (Commutative Law of Multiplication):

$$xy = yx,$$

*where x and y are variables whose domains are the set of cardinal numbers.*

*Note*: The meaning of this statement is the endless list of statements $0 \cdot 0 = 0 \cdot 0$ and $0 \cdot 1 = 1 \cdot 0$ and $\cdots$ and $55 \cdot 97 = 97 \cdot 55$ and $\cdots$.

*Proof*: We need only show that if $a$ and $b$ are any cardinal numbers then $ab = ba$. Let $A$ and $B$ be sets having cardinal numbers $a$ and $b$ respectively. Then, by definition of multiplication,

$ab$ is the cardinal number of $A \times B$,
$ba$ is the cardinal number of $B \times A$.

Since $A \times B$ and $B \times A$ can be put into one-to-one correspondence (it is obvious how this can be done), $ab$ and $ba$ are the same.

This theorem is known as the *Commutative Law of Multiplication*. We shall refer to it as CM.

Theorem IV (Associative Law of Multiplication): *Let $(ab)c$ designate the product whose first factor is $ab$ and whose second factor is $c$. Let $a(bc)$ designate the product whose first factor is $a$ and whose second factor is $bc$. Then*

$$(xy)z = x(yz),$$

*where x, y, and z are variables whose domains are the set of all cardinal numbers.*

*Note*: This theorem asserts the truth of an endless list of statements, such as $(2 \cdot 3)5 = 2(3 \cdot 5)$.

*Proof*: Let $A, B, C$ be sets whose respective cardinal numbers are $a, b, c$. Then, by definition of multiplication,

$(ab)c$ is the cardinal number of $(A \times B) \times C$
$a(bc)$ is the cardinal number of $A \times (B \times C)$.

We are going to show that the sets $(A \times B) \times C$ and $A \times (B \times C)$ can be put into one-to-one correspondence, from which it follows that $(ab)c$ and $a(bc)$ are the same. To that end, take any member of $(A \times B) \times C$ and observe that it is an ordered pair whose first member is a member of $A \times B$ and whose second is a member of $C$. Thus, each member of $(A \times B) \times C$ can be described by a symbol $((p, q), r)$, where $p$ is a member of $A$, $q$ is a member of $B$, and $r$ is a member of $C$. It is formed from a member of $A$, a member of $B$, and a member of $C$. Similarly, each member of $A \times (B \times C)$ can be described by a symbol $(p, (q, r))$. The required one-to-one correspondence between the sets can be found by pairing off $((p, q), r)$ of $(A \times B) \times C$ with $(p, (q, r))$ of set $A \times (B \times C)$.

This law is known as the *Associative Law of Multiplication*. We shall refer to it as Law AM.

## The Distributive Law

Each of the four laws already established is concerned exclusively with addition or with multiplication. We prove a theorem in which multiplications and additions figure jointly.

THEOREM V: *Let $a(b + c)$ designate the product whose first factor is $a$ and whose second factor is $b + c$. Let $(ab) + (ac)$ designate the sum whose first term is $ab$ and whose second term is $ac$. Then*

$$x(y + z) = (xy) + (xz),$$

*where $x$, $y$, and $z$ are variables whose domains are the set of all cardinal numbers.*

*Note*: This theorem asserts the truth of an endless list of statements, such as $2(3 + 5) = (2 \cdot 3) + (2 \cdot 5)$.

*Proof*: Let $A$, $B$, and $C$ be sets having the respective cardinal numbers $a$, $b$, and $c$, and suppose that $B$ and $C$ have no members in common. Let $S$ denote the union of $B$ and $C$. Then, by our definitions of multiplication and addition

$a(b + c)$    is the cardinal number of $A \times S$;
$(ab) + (ac)$ is the cardinal number of the union of $A \times B$ and
$A \times C$.

It follows directly from the definition of union and of product set
that $A \times S$ is the same set as the union of $A \times B$ and $A \times C$.
Hence, $a(b + c) = (ab) + (ac)$.

This theorem is known as the *Distributive Law*; we shall refer
to it as Law D.

## Comments on the Five Basic Laws

These laws have two quite different roles in our development
of algebra. In the first place, they will be used as a basis for
deriving most of the important properties of the system of cardinal
numbers. They will also figure prominently in our construction
of other, more complicated, number systems. Each of the five basic
laws states that a certain statement-form becomes a true statement
if the domain of the variables is specified to be the set of cardinal
numbers. Later we shall encounter new sets of entities, by no
means as familiar as the cardinals, that we shall also call number
systems. Our recognition of these new sets as number systems will
be due in part to the fact that these same five statement-forms apply.
While we cannot give here a final definition of the idea "number
system," we can stipulate that no set $S$ will be considered as a
number system if the statement-forms contained in the five basic
laws do not become true statements when the variables in them
have domain $S$.

## Cancellation Laws

We now state two theorems about the cardinal number system
whose truth we shall assume here without proof. They are:

THEOREM VI (Cancellation Law for Addition): *If $a$, $b$, and $c$
are cardinals for which*

$$a + b = a + c,$$

*then*

$$b = c.$$

Theorem VII (Cancellation Law for Multiplication): *If $a$, $b$, and $c$ are cardinals for which $ab = ac$ and if $a$ is not zero, then $b = c$.*

We shall use these theorems about cardinals without proof. Their proofs are taken up in the appendix as part of the discussion of infinite sets.

## Special Properties of Zero and One

The cardinal number zero has two special properties which we now state and prove. They are:

Theorem VIII:
$$0 + x = x$$
$$0 \cdot x = 0,$$

*where $x$ is a variable whose domain is the set of all cardinal numbers.*

*Proof:* Let $a$ be any cardinal, let $A$ be a set whose cardinal number is $a$, and let $E$ be the empty set. Then $a + 0$ is, by definition, the cardinal number of the union of $A$ and $E$. But this union is $A$ itself, whose cardinal number is $a$. Thus $a$ and $a + 0$ are the cardinal numbers of the same set and hence are the same cardinal number. This proves the first statement. To prove the second, construct $E \times A$, whose cardinal number is, by definition, $0 \cdot a$. We are to show that the cardinal number of $E \times A$ is 0. We know that $E \times A$ consists of ordered pairs whose first entries come from $E$ and whose second entries come from $A$. But $E$ is the empty set and can furnish no such entries. It follows that $E \times A$ has no members, that it is the empty set, and that its cardinal number is 0.

Theorem IX:
$$x \cdot 1 = x,$$

*where $x$ is a variable whose domain is the set of all cardinal numbers.*

*Proof:* Let $a$ be any cardinal number; let $A$ be a set whose cardinal number is $a$ and whose members are $p$, $q$, $r$, $\ldots$; let $I$ be a set with cardinal number 1 whose single member is $\#$. Then $A \times I$ has the members $(p, \#)$, $(q, \#)$, $(r, \#)$, $\cdots$. To prove that $a \cdot 1 = a$, we observe that $A$ and $A \times I$ can be put in one-to-one correspondence by pairing $p$ with $(p, \#)$, $q$ with $(q, \#)$, $r$ with $(r, \#)$, and so forth. Then $A \times I$ and $A$ have the same cardinal number. But the cardinal number of $A$ is $a$, and that of $A \times I$ is $a \cdot 1$. This proves our theorem.

We have proved that zero has two special properties. We now show that zero is the only cardinal that has either of these properties. More precisely we show

Theorem X: *Let $u$ and $a$ be cardinals.*

   (i) *If $u + a = a$, then $u = 0$;*

   (ii) *if $u \cdot a = u$ and $a$ is not $1$, then $u = 0$.*

To prove (i) we have

$$u + a = a \qquad \text{(given)}$$
$$= 0 + a \qquad \text{(Theorem VIII)}$$

whence
$$u = 0 \qquad \text{(Theorem VI)}.$$

To prove (ii) we have

$$u \cdot a = u \qquad \text{(given)}$$
$$= u \cdot 1 \qquad \text{(Theorem IX)}.$$

Then, if $u$ were not zero, we would have

$$a = 1 \qquad \text{(Theorem VII)}$$
$$a \text{ is not } 1 \qquad \text{(given)}.$$

Then, to avoid a contradiction, we must have $u = 0$.

The following theorem is frequently used:

Theorem XI: *If $a$ and $b$ are cardinals, and if*

$$ab = 0$$

*then either $a = 0$ or $b = 0$.*

*Proof:* If $a$ is not zero, we have

$$ab = 0 \qquad \text{(given)}$$
$$= a \cdot 0 \qquad \text{(Theorem VIII)}$$
$$b = 0 \qquad \text{(Theorem VII)}.$$

Clearly, having shown that if $a$ is not zero, $b$ must be zero, we have proved that at least one of the two numbers $a$ and $b$ must be zero.

We now show that the property of the number one given in Theorem IX characterizes it. Specifically,

THEOREM XII: *If $u$ and $a$ are cardinals for which $a \cdot u = a$, and if $a$ is not zero, then $u = 1$.*

*Proof:*

$$a \cdot u = a \qquad \text{(given)}$$
$$= a \cdot 1 \qquad \text{(Theorem IX)}$$
$$u = 1 \qquad \text{(Theorem VII).}$$

## The Addition and Multiplication Tables for Cardinals

We are now able to tackle such problems as, How much is $2 + 2$? We recognize that the reader can supply the answers for the numbers of everyday life, having learned them by rote. We hope that he will not be distressed by the resemblance between what we prove and what he was told in second grade.

Let us first formulate the problem precisely. The cardinal numbers $a$ and $b$ being given, their sum $a + b$, and their product $ab$ are cardinal numbers and must occur in the list $0, 1, 2, 3, \cdots$. The problem is to find a procedure for determining what member of the list is designated by such a sum or product. This is the problem of finding the standard form for a sum or a product of cardinals.

We already have some information in the case of addition. We have in fact solved this problem in the case in which one of the terms is $0$. Let us examine the case in which one of the terms is $1$. Recall that the standard set which is the immediate successor of a standard set $q$ is the union of $q$ with the set whose only member is $q$. Thus, the cardinal number of the immediate successor of $q$ is, by definition of addition, the sum of the cardinal numbers of these two sets. Now, the cardinal number of the standard set $q$ is $q$, the cardinal number of the set whose only member is $q$ is $1$. Thus the cardinal number of the immediate successor of $q$ is $q + 1$. We can thus state: $0 + 1 = 1$, $1 + 1 = 2$, $2 + 1 = 3$, $3 + 1 = 4, \cdots$. In statement-form variable language, we have shown that the successor of $x$ is $x + 1$, where $x$ is a variable whose domain is the set of cardinals.

Now let us examine a problem which involves neither $0$ nor $1$. Let us try to find the standard form for $2 + 2$. We have

$$2 + 2 = 2 + (1 + 1) \qquad \text{(because } (1+1)=2)$$
$$= (2+1)+1 \qquad \text{(AA)}$$
$$= 3+1 \qquad (2+1=3)$$
$$= 4 \qquad (3+1=4).$$

The first step in this argument rested on our knowing that $2 = 1 + 1$. The reader may concede that we showed earlier that $1 + 1 = 2$, but may feel that we have somehow used this information unfairly by reversing the order. We can only remind him that the statement $1 + 1 = 2$ means that $1 + 1$ is the same number as 2 and that this in turn implies that 2 is the same number as $1 + 1$. The remaining steps are straightforward.

A slight modification of this argument gives a procedure for finding the standard form for any sum one of whose terms is 2, say $a + 2$. We have

$$a + 2 = a + (1 + 1) \qquad \text{(because } 2 = 1 + 1)$$
$$= (a+1)+1 \qquad \text{(AA)}.$$

The sum $a + 1$ can be written in standard form when $a$ is specified; and then the standard form for $(a + 1) + 1$ is known. For instance

$$17 + 2 = 17 + (1 + 1) \qquad (2 = 1 + 1)$$
$$= (17+1)+1 \qquad \text{(AA)}$$
$$= 18+1 \qquad (17+1=18)$$
$$= 19 \qquad (18+1=19).$$

We handled the term 2 by reducing it to several applications of our knowledge of the term 1. The term 3 can also be referred back similarly. Assuming the procedure for treating the term 2 as known, let us attempt a problem involving 3—for example, $26 + 3$. We have

$$26 + 3 = 26 + (2 + 1) \qquad (3 = 2 + 1)$$
$$= (26+2)+1 \qquad \text{(AA)}$$
$$= 28+1 \qquad (26+2=28)$$
$$= 29 \qquad (28+1=29).$$

More generally,

$$a + 3 = a + (2 + 1) \qquad (3 = 2 + 1)$$
$$= (a+2)+1 \qquad \text{(AA)};$$

and if the standard form for $a + 2$ can be found, so can the standard form for $(a + 2) + 1$. In fact, if we can manage successfully the term $b$, we can also manage $b + 1$ as a term, because $a + (b + 1) = (a + b) + 1$ and the latter involves the term $b$ and then the term 1. Thus, once 0 and 1 become tractable, so do $1 + 1 = 2$, $2 + 1 = 3$, $3 + 1 = 4$, ..., and all cardinals.

Standard forms for products can be treated by a similar process of reduction. For products with a factor 0 or 1, the problem is solved by earlier work. Let us turn to $2 \cdot 2$. We have

$$
\begin{aligned}
2 \cdot 2 &= 2(1 + 1) & &(2 = 1 + 1) \\
&= (2 \cdot 1) + (2 \cdot 1) & &(D) \\
&= 2 + 2 & &(2 \cdot 1 = 2) \\
&= 4 & &(2 + 2 = 4).
\end{aligned}
$$

More generally,

$$
\begin{aligned}
a \cdot 2 &= a(1 + 1) & &(2 = 1 + 1) \\
&= (a \cdot 1) + (a \cdot 1) & &(D) \\
&= a + a & &(a \cdot 1 = a),
\end{aligned}
$$

and the problem is referred back to addition.

To handle 3 as a factor, we can write

$$
\begin{aligned}
26 \cdot 3 &= 26(2 + 1) & &(3 = 2 + 1) \\
&= (26 \cdot 2) + (26 \cdot 1) & &(D) \\
&= (26 \cdot 2) + 26 & &(26 \cdot 1 = 26),
\end{aligned}
$$

and if it is known that $26 \cdot 2 = 52$, we have referred the whole question back to the already solved case of addition. The procedure can also be applied to 4 as a factor, and to 5, 6, ..., and each of the cardinals.

## Exercises

1. Starting with some familiar set $A$ whose cardinal number is 3 and some familiar set $B$ whose cardinal number is 2, construct a set whose cardinal number is

a) $2 + 3$;

b) $3 + 2$;

c) $2 + (3 + 2)$;

d) $(2 + 3) + 2$;

e) $2 \cdot 3$;

f) $3 \cdot 2$;

g) $2(3 \cdot 2)$;

h) $(2 \cdot 3)2$;

i) $2(3 + 2)$;

j) $(2 \cdot 3) + (2 \cdot 2)$.

2. Prove the following, using the five laws but not using the addition or multiplication tables ($a$ is supposed to denote a cardinal number):

a) $2(3 + 4) = 2(4 + 3)$.

b) $2 + (3 + 4) = 4 + (2 + 3)$.

c) $5(3 \cdot a) = a(5 \cdot 3)$.

d) $5(3 + a) = (a + 3)5$.

e) $5(3 + a) = (5 \cdot a) + (3 \cdot 5)$.

3. Prove the following, using the cancellation laws and the addition and multiplication tables ($a$ is supposed to denote a cardinal number):

a) If $3 + 5 = 3 + a$, then $5 = a$.

b) If $8 = 3 + a$, then $5 = a$.

c) If $7a = 21$, then $a = 3$.

d) If $7 + a = 21$, then $a = 14$.

e) If $2(5 + a) = 20$, then $a = 5$.

f) If $7 \cdot a = 0$, then $a = 0$.

g) There is no cardinal number $a$ for which $a = a + 1$.

h) There is no cardinal number $a$ for which $2a = 1$.

# CHAPTER III

## EXPRESSIONS

### Introduction

THE BASIC LAWS ALREADY DISCUSSED constitute a collection of five statements with five individual and different meanings. It should be clear, however, that they resemble one another; that they are all the same kind of statement. In this chapter we investigate a vast body of similar statements and show how they may be formed and how they may be verified. This analysis is one of the principal activities of algebra.

### Parentheses

We have seen that the operations of addition and multiplication, defined originally for two cardinal numbers, may be used step by step to combine three cardinal numbers. In a similar way, we can use the operations to combine four, five, or more cardinal numbers, the only restriction being that we must not attempt to combine more than two cardinal numbers in any single step. As before, a judicious use of parentheses enables us to record the result of such repeated operations. Before stating the general principle, we consider some examples.

*Examples*: $(2 + 7) + (5 + 3)$ designates the sum whose first term is itself a sum $2 + 7$ and whose second term is likewise a sum $5 + 3$. $(2 + 7)(5 + 3)$ designates the product whose first factor is the sum $2 + 7$ and whose second factor is the sum $5 + 3$.

In both these examples, the parentheses enclose a single number. It might seem that the parentheses in $(2 + 7)$ enclose two numbers and a plus sign, but our definition of addition and our agreement on the use of the plus sign require us to consider the combination of symbols $2 + 7$ as a way of designating a single number. For symbols in standard form, no problem arises. Thus we write—with no parentheses—the sum $9 + 8$ and the product $9 \cdot 8$.

In general, if we have any combination of symbols that designates a single number, we can, by enclosing that combination in parentheses, use it as a term in a sum, or as a factor in a product.

## Expressions

We define the useful concept *expression* by listing the things that are to be so described.

1) Each cardinal number in standard form is an expression;
2) each variable whose domain is the set of cardinal numbers is an expression;
3) if $A$ and $B$ are expressions, then $(A) + (B)$ is an expression;
4) if $A$ and $B$ are expressions, then $(A)(B)$ is an expression.

*Note*: Items 3) and 4) state that any two expressions can be used to form new expressions by enclosing each in parentheses and then combining them by either addition or multiplication. We attach a further stipulation that if either of the expressions so combined is a cardinal number in standard form or a variable whose domain is the set of all cardinal numbers then no parentheses are to enclose that expression.

*Examples*: Items 3) and 4) certainly certify as expressions each of the following: $3 + 5$, $3 \cdot 5$, $3 + x$, $3x$, $x + y$, $xy$. They also certify as expressions more complicated combinations such as

$$(3 + 5) + (6 + 11),$$
$$(3 + x) + (yz),$$
$$((x + y)(5z)) + (3w).$$

Let us examine the last of these. Clearly $x + y$ and $5z$ are expressions, and so, by item 4), we can conclude that $(x + y)(5z)$ is an expression. Moreover, this last expression can figure in the construction of new expressions, and if we combine it with the expression $3w$ in accordance with item 3), we do in fact find that

$$((x+y)(5z))+(3w)$$

is an expression.

## Evaluation of Expressions

Every expression has this important property: If each variable in it is replaced by a cardinal number, then the combination of symbols thus obtained designates a cardinal number. While this is true no matter how the replacements are made, let us agree to study only substitutions in which a given variable is replaced by the same number in each place in the expression where that variable occurs. If we have an expression constructed from certain variables, we shall frequently effect such replacements of the variables. We shall call the numbers that replace the variables the *values of the variables*, and we shall call the number designated by the new combination of symbols the *value of the expression for the given values of the variables*. Once values for the variables are chosen, the value of the expression is a definite number, but the selection of other values for the variables might yield a different value for the expression. There is no way of determining, merely by inspecting an expression, which values of the variables are to be used, or indeed if any are to be used. It should simply be noted that if, for any reason, values are chosen for the variables, then the expression itself has a value. It should be further noted that any expression will define a function if we agree to assign to each member of the set of possible values for the variables the corresponding value of that expression.

*Example*: The value of the expression $((x+y)(5x))+(3y)$ for the values 3 for the variable $x$ and 2 for the variable $y$ is the number $((3+2)(5 \cdot 3))+(3 \cdot 2)$. If we denote the given expression by $F(x,y)$, we write

$$F(x,y)=((x+y)(5x))+(3y)$$

and

$$F(3,2)=((3+2)(5 \cdot 3))+(3 \cdot 2).$$

This number is found, by use of the tables, to be 81. The set of values of a given expression is called its *range*. This set may consist of the whole set of cardinal numbers (as in the case of the expression $xy$) or of a proper subset of the set of cardinals (as in the case of the expression $2x$, whose range is the even cardinals).

## Equal Expressions

Let $E$ be an expression constructed from certain cardinal numbers and certain variables. Let $F$ be another expression constructed from these same variables (but not necessarily from these same cardinals). We shall say that $E = F$ (read: $E$ equals $F$) if for all possible replacements of the variables by cardinal numbers the value of $E$ is the same as the value of $F$. Thus we are, in effect, using the statement-form variable language introduced in Chapter I.

*Example*: Let $E$ be the expression $x(y + z)$ and let $F$ be the expression $(xy) + (xz)$. Then we assert that

$$x(y + z) = (xy) + (xz).$$

Our authority for this statement is Law D.

At the beginning of this chapter we referred to a vast body of statements whose discovery is one of the principal concerns of our science. We can now describe more specifically the nature of those statements. They are all statements that one expression equals another. We cannot hope to discover all such statements, but we do propose to show

 a) how to construct new expressions equal to a given one;
 b) how, from pairs of equal expressions, to construct other pairs of equal expressions;
 c) how, given two expressions, to determine if they are equal.

We describe these processes as "the algebra of expressions."

## The Algebra of Expressions

Any expression other than a number in standard form, or a variable, is constructed as a combination of other expressions, which we shall call its *components*.

*Example*: $(x + y)(3z)$ has the components $x, y, x + y, 3, z, 3z$.

We now state and prove a theorem that enables us to form new expressions equal to a given one.

THEOREM I: *Let E be an expression; let C be a component of E. Let C′ be an expression that is known to be equal to C. Let E′ be the expression obtained by replacing C by C′ in E. Then, E = E′.*

*Example*: Let $E$ be $x(y + z)$; let $C$ be $y + z$; and let $C′$ be $z + y$ (known to be equal to $C$ by law CA). Then $E′$ is $x(z + y)$ and the theorem claims that

$$x(y + z) = x(z + y)$$

*Proof*: To prove the statement $E = E′$, we must show that no matter what values are given to the variables in $E$ and $E′$, the values of $E$ and $E′$ will be the same. If values are chosen for the variables, the procedure for finding the value of $E$ differs from that for finding the value of $E′$ in only one respect: wherever, in the former, we use the value of $C$, in the latter we use the value of $C′$. But $C = C′$, so these values must be the same, and thus, after the values for $C$ and $C′$ are found, the two procedures are identical and must lead to the same result. We next state and prove a theorem that enables us to derive new pairs of equal expressions from a given pair of equal expressions.

THEOREM II: *Let E and F be equal expressions involving certain variables $x, y, \ldots, w$. Let $X, Y, \ldots, W$ be any expressions. Let E′ be the expression obtained by replacing in E the variable x by the expression X, the variable y by the expression Y, . . . , the variable w by the expression W; and let F′ be obtained similarly from F. Then, E′ = F′.*

*Example*: Let $E$ be $x + y$; $F$ be $y + x$; $X$ be $pq$; and $Y$ be $q$. Then, $E′$ is $(pq) + q$ and $F′$ is $q + (pq)$, and the theorem claims that

$$(pq) + q = q + (pq).$$

*Proof*: We must show that no matter what values are given to the variables in $E′$ and $F′$, the values of $E′$ and $F′$ are the same. When we choose values for the variables that appear in the expressions $X, Y, \ldots, W$, these expressions acquire values, and the values of $E'$ and $F'$ are expressed in terms of these values of the expressions $X, Y, \ldots, W$. But the very same combinations of values can also be obtained by assigning to the variables $x, y, \ldots, w$ in $E$ and $F$ the values of $X, Y, \ldots, W$ respectively, and we know that the values of $E$ and $F$ thus obtained are the same. This proves that the values of $E'$ and $F'$ must be the same; hence $E' = F'$.

Corollary: *The five fundamental laws hold for expressions* (as well as for variables).

A much simpler theorem about equal expressions is

Theorem III: *If $E$, $F$, and $G$ are expressions for which $E = F$ and $F = G$, then $E = G$.*

*Proof*: When values for the variables are chosen, $E$, $F$, and $G$ acquire values; we are to show that the value of $E$ must be the same as the value of $G$. We know that the value of $E$ is the same as the value of $F$, because $E = F$, and we know that the value of $F$ is the same as the value of $G$, because $F = G$. Thus the value of $E$ and the value of $G$ are same, namely, the value of $F$.

## Parentheses and the Generalized Addition Law

Most of the expressions with which we have dealt contain many more parentheses than are customarily used. We now take up the problem of eliminating the extra ones. The reader who wonders why they were introduced in the first place if they are to be removed, is invited to observe that their role is carefully preserved, and that it is only by making additional agreements about our symbols that we make it possible to leave out certain of the parentheses.

We make the convention about omitting parentheses that $x + y + z$ (up till now a meaningless array of symbols) shall mean $x + (y + z)$. We understand that this convention applies to other arrangements of the variables (for instance, $y + x + z$ means $y + (x + z)$) and to other variables (for instance, $p + q + r$ means $p + (q + r)$). Thus, with the three variables $x, y, z$ we can construct two lists of expressions:

|            List I            |                   |    List II    |
| :--------------------------: | :---------------: | :-----------: |
| $x + (y + z)$                | $(x + y) + z$     | $x + y + z$   |
| $x + (z + y)$                | $(x + z) + y$     | $x + z + y$   |
| $y + (x + z)$                | $(y + x) + z$     | $y + x + z$   |
| $y + (z + x)$                | $(y + z) + x$     | $y + z + x$   |
| $z + (x + y)$                | $(z + x) + y$     | $z + x + y$   |
| $z + (y + x)$                | $(z + y) + x$     | $z + y + x$   |

It is fairly easy to prove that every expression in List I is equal to every other expression in that list. Let us prove, for example, that

$$(x + y) + z = y + (z + x).$$

We have

$(x + y) + z = z + (x + y)$     (by CA and Theorem II)
$z + (x + y) = (z + x) + y$     (by AA)
$(z + x) + y = y + (z + x)$     (by CA and Theorem II).

Then, by Theorem III, the first expression of this chain equals the last. Each of the other equalities can be verified similarly. Let us suppose this done. Since each expression of List II is, by definition, equal to an expression of List I, we then have, by Theorem III that each expression of either list is equal to every other expression in either list.

When an expression $x + (y + z)$ appears as a component in another expression, it may still, of course, be written in the form $x + y + z$. Thus, the expression $w + (x + (y + z))$ is equal to $w + (x + y + z)$. The same kind of analysis applies to expressions with four, five, or any number of variables. We define, successively,

$$x + y + z + w \text{ to mean } x + (y + z + w),$$
$$x + y + z + w + t \text{ to mean } x + (y + z + w + t),$$
$$\text{etc.}$$

We then observe that an expression formed from any number of variables by repeated additions is equal to the expression obtained from it by the removal of all its parentheses. For instance, $(x + y) + (z + t)$ is equal to $x + y + z + t$. (One way to prove this in detail is by proving $(x + y) + (z + t) = x + (y + z + t)$ and then applying the definition of $x + y + z + t$.) We next observe that if the terms in a sum without parentheses are rearranged, the resulting expression is equal to the original. For instance, $x + y + z + t + w$ is equal to $t + z + y + w + x$. We shall refer to these facts as the *General Addition Law* (GA). GA guarantees that if from an expression that involves only additions we form a new expression by inserting or removing parentheses or by rearranging terms, then the new expression will be equal to the original one.

It should be mentioned explicitly that while we were chiefly concerned in this section with parentheses, we have also added to our information about equal expressions.

## Parentheses and the Generalized Multiplication Law

We follow the aims and methods of our study of the use of parentheses in addition. We define $xyz$ (up till now meaningless) to be $x(yz)$. We understand this definition to apply to other arrangements of these variables, and to other variables; that is, $yzx$ means $y(zx)$ and $pqr$ means $p(qr)$. We can then list all the expressions that can be formed from the three variables $x$, $y$, and $z$ by multiplications without repetitions.

| | | |
|---|---|---|
| $x(yz)$ | $(xy)z$ | $xyz$ |
| $x(zy)$ | $(xz)y$ | $xzy$ |
| $y(xz)$ | $(yx)z$ | $yxz$ |
| $y(zx)$ | $(yz)x$ | $yzx$ |
| $z(xy)$ | $(zx)y$ | $zxy$ |
| $z(yx)$ | $(zy)x$ | $zyx$ |

It can be shown readily that each expression on this list is equal to every other one. In particular, if we insert parentheses in an expression on this list or remove them, or if we rearrange the factors of the expression, the new expression thus obtained is also on this list and is therefore equal to the old one. Similar facts hold for expressions involving four, five, or any number of variables. We first define, successively,

$$xyzw \text{ to mean } x(yzw)$$
$$xyzwt \text{ to mean } x(yzwt)$$
$$\text{etc.}$$

We then observe that an expression that is formed from any number of variables by repeated multiplications is equal to the expression that is obtained from it by the removal of all its parentheses (for instance, $(xy)(zt)$ is equal to $xyzt$). Next, we observe that if the factors in a product involving no parentheses are rearranged, the new expression is equal to the old (for instance, $xyztw$ is equal to $txywz$). We shall refer to these facts as the *Generalized Multiplication Law* (GM). GM guarantees that if, from an expression that involves only multiplications, we form a new expression by removing or inserting parentheses or by rearranging the factors, then the new expression will be equal to the original.

## Parentheses and the Generalized Distributive Law

Our discussion in each of the two preceding sections has resulted in the removal of parentheses in certain cases. We now give an example in which the removal of parentheses would lead to ambiguity. It involves both addition and multiplication. Consider the following combinations of symbols: $(2 + 7)3$, $2 + (7 \cdot 3)$, $2 + 7 \cdot 3$. The first of these equals $9 \cdot 3$, or 27, the second is $2 + 21$, or 23. The third is not an expression at all but can be obtained from each of the first two by omitting parentheses. Thus if we were to omit parentheses indiscriminately, we would in this case be using the same symbol for 23 and for 27—an undesirable ambiguity. There are, however, only two meanings that $2 + 7 \cdot 3$ might reasonably have, namely, $(2 + 7)3$, and $2 + (7 \cdot 3)$. We arbitrarily choose one of these meanings, avoiding ambiguity and yet removing parentheses. We agree that $2 + 7 \cdot 3$, meaningless till now, shall mean $2 + (7 \cdot 3)$. More generally, for any three expressions, $A, B, C$, we agree that $A + BC$ shall mean $(A) + (BC)$. In other words, when no parentheses appear in an expression that involves multiplications and additions, we shall consider the expression to be a sum of products rather than a product of sums. We use this convention in formulating the following important result.

Theorem IV: *Let $p, q, \ldots, s$ and $x, y, \ldots, w$ be variables whose domains are the set of cardinal numbers. Then*

$$(p + q + \ldots + s)(x + y + \ldots + w) = px + py + \ldots + pw$$
$$+ qx + qy + \ldots + qw + \ldots + sx + sy + \ldots + sw.$$

*Note*: This theorem tells us how to convert a product of sums into a sum of products. The dots are supposed to convey the fact that the number of variables in each sum is not restricted, that the theorem is true no matter how many variables appear in each sum. The case in which the first factor has one term and the second two terms should be recognizable as Law D. What this theorem accomplishes is to generalize Law D, and it is known as the Generalized Distributive Law (GD).

*Proof*: We prove the theorem by cases, considering first the case in which the first factor has one term and the second three. We are to prove that

$$p(x + y + z) = px + py + pz.$$

We have

$$p(x + y + z) = p(x + (y + z)) \qquad \text{(by GA)}$$
$$p(x + (y + z)) = px + p(y + z) \qquad \text{(by the corollary to Theorem II applied to D)}$$

$$px + p(y + z) = px + py + pz \qquad \text{(by D)}$$
$$p(x + y + z) = px + py + pz \qquad \text{(by Theorem III)}.$$

In our next case, the first factor has one term and the second four. We are to prove that

$$p(x + y + z + w) = px + py + pz + pw.$$

We have

$$p(x + y + z + w) = p(x + (y + z + w)) \qquad \text{(by GA)}$$
$$p(x + (y + z + w)) = px + p(y + z + w) \qquad \text{(by the corollary to Theorem II applied to D)}$$

$$px + p(y + z + w) = px + py + pz + pw \quad \text{(by the previous case)}$$

$$p(x + y + z + w) = px + py + pz + pw \quad \text{(by Theorem III)}.$$

Similarly, we could prove the theorem for the case in which the first factor has one term and the second five or six or any number of terms. Let us suppose this done, and consider next the case in which the first factor has two terms, the second any number. We are to prove that

$$(p + q)(x + y + \ldots + w)$$
$$= px + py + \ldots + pw + qx + qy + \ldots + qw.$$

We have

$$(p + q)(x + y + \ldots + w)$$
$$= (x + y + \ldots + w)(p + q)$$

$$\text{(by the corollary to Theorem II applied to CM)}$$

$$= (x + y + \ldots + w)p + (x + y + \ldots + w)q$$

$$\text{(by the corollary to Theorem II applied to D)}$$

$$= p(x + y + \ldots + w) + q(x + y + \ldots + w)$$

$$\text{(by the corollary to Theorem II applied to CM)}$$

$$= px + py + \ldots + pw + qx + qy + \ldots + qw$$

$$\text{(by the previous case)}.$$

Similarly, a proof could be constructed for the case in which the first factor has three terms, four terms, or any number of terms. This completes our proof of this theorem.

The three laws GA, GM, GD are theorems that generalize the five basic laws. Each of these theorems is a statement asserting that one expression equals another expression. The five basic laws are also such statements, but the five basic laws refer to five specific pairs of expressions, whereas the three generalized laws refer to an endless list of pairs of expressions. In the introduction to this chapter, we proposed to derive "a vast body of statements." In proving the three generalized laws, we have carried out this task.

## EXERCISES

1. List all the expressions that can be formed from the number 5 and the letter $x$:

    a) using one operational symbol (8 expressions);

    b) using two operational symbols (64 expressions).

2. a) Sort the expressions of 1a) above into sets of equal ones.

    b) Sort the expressions of 1b) above into sets of equal ones.

3. Prove that each of the following is an expression:

$$x + (y + (z + w)), \quad (x + y) + (z + w), \quad x(3 + 4),$$
$$x(y(zw)), \quad (xy)(zw), \quad (x + 3)(4 + 5),$$
$$((2x) + 3)((4y) + 8).$$

4. a) By a single application of CA, obtain from $x + (y + z)$ a new expression that is equal to it    (2 answers).

    b) The same for $(x + y) + (y + z)$    (3 answers).

    c) The same for $x + (y + z)$ and AA    (1 answer).

    d) The same for $(x + y) + (y + z)$ and AA

                                           (2 answers).

5. a) The same as 4a) for $x(yz)$ and CM    (2 answers).

    b) The same for $(xy)(zw)$ and CM    (3 answers).

    c) The same for $x(yz)$ and AM    (1 answer).

    d) The same for $(xy)(zw)$ and AM    (2 answers).

6. a) Given $F(x) = (x + 2)(x + 3)$, state what is meant by $F(1)$, $F(3)$, $F(2x)$, $F(x + 3)$.

    b) Given $F(x, y) = ((3x + 4y)(y + 3)) + (2x)$, state what is meant by $F(1, 2)$, $F(3, 5)$, $F(3, y)$, $F(x, 4)$.

7. Apply GD to each of the following:

$$(x + 2)(x + 3).$$
$$x(x + y + 3).$$
$$(x + y + z)(2 + 4 + z).$$

8. Find the standard form for each of the following:

a) $1(3x + 4)$.
b) $0(3x + 4)$.
c) $(0 + 1)(3x + 4)$.
d) $(3 + 4)x + 1$.
e) $1 + x^{3+4}$
f) $(x^2)^3 + 1$.
g) $x^{(2^3)} + 1$.

# CHAPTER IV

## POLYNOMIALS

### Introduction

IN THIS CHAPTER, WE shall construct a test for the equality of two expressions. We regard the construction of such a test as an important part of algebra, because the statements which make up this subject are mostly statements that one expression equals another. The test is easy to outline. We shall find, for each expression, an equal expression of a special kind, which we shall call its standard form. It will be shown that equal expressions have the same standard form, and that unequal expressions have different standard forms. The test to determine whether or not two given expressions are equal is then simply to find the standard form of each and compare them.

### Expressions with No Parentheses

Our next theorem goes as far as it is possible to go in the removal of parentheses. It asserts the possibility of replacing any expression by an equal one that has no parentheses at all. Before we take up the details, some general remarks should be made. It should be stated emphatically that there is nothing undesirable about parentheses. Our efforts at removing them are directed toward devising a test to determine if two given expressions are equal. The following theorem is a crucial detail of that attempt.

THEOREM I: *For each expression E there is an equal expression F which has no parentheses.*

*Proof:* An expression must contain either variables, cardinal numbers in standard form, or both. Let $E$ be an expression, and let $a, b, \ldots, f$ be the cardinal numbers in standard form and $u, v, \ldots, z$ the variables that are used in its construction. If $E$ happens to consist of a single variable or of a single cardinal in standard form, there is nothing to prove, for $E$ has no parentheses. Otherwise, the first step in the construction of $E$ must have been the selecting of two expressions from the set $a, b, \ldots, f$; $u, v, \ldots, z$ and then either adding or multiplying them. If this first step produced $E$, there is again nothing to prove, for the result of such a step would involve no parentheses. If this step did not produce $E$, call the result of the step $P$. Then $P$ is a component of $E$. The next step in the construction of $E$ must have been the selecting of two expressions from the set $a, b, \ldots, f$; $u, v, \ldots, z$; $P$ and either adding or multiplying them. The expression obtained at this step involves parentheses only if $P$ was one of the two expressions selected. In this case it has one of six possible forms: $t(P)$, $(P)t$, $(P)(P)$, $(P)+t$, $t+(P)$, $(P)+(P)$, where $t$ is one of $a, b, \ldots, f$; $u, v, \ldots, z$. We now show that for each of these expressions we can construct one that is equal to it and free of parentheses. If $P$ is a product, say $rs$, then GM allows us to write the first three of the above expressions without parentheses, and our agreement about sums of products allows the removal of those that figure in the last three. If $P$ is a sum, then GA allows the removal of the parentheses in the last three of the above expressions, and GD allows their removal in the first three. Thus the expression produced by this second step has an equal one with no parentheses. If this second step produced $E$, we have our desired result. If not, call the result of this second step $Q$, and observe that the third step consists in adding or multiplying some two of $a, b, \ldots, f$; $u, v, \ldots, z$; $P$, $Q$. (Actually, instead of $Q$ itself an expression equal to $Q$ may have been used in the formation of $E$. However, we know that if we substitute for one component an equal one, the new expression equals the original one.) Since none of these involve parentheses, we can show, by reasoning as before, that a sum or product of two of them equals an expression with no parentheses. We see, by continuing in this way, that at each step an expression with no parentheses is obtained. The final step yields an expression that is equal to $E$ and that contains no parentheses.

## Exponents

Exponents first appeared in mathematics in the 14th century, when they were used just as abbreviations, and they gradually acquired a more profound role. In this chapter we discuss them as they were originally used. We also discuss exponents in Chapter XI, where their more recent meaning is given.

DEFINITION: Let $E$ be an expression and let $n$ be a cardinal number different from zero. We shall mean by $E^n$ (read: $E$ to the $n$-th power) the expression

$$E \cdot E \cdot \ldots \cdot E,$$

where $E$ appears as a factor $n$ times. The cardinal number $n$ in $E^n$ is called the *exponent* of $E$.

*Examples*:  $3^2 = 9$,
$$5^3 = 125,$$
$$8^{14} = 8 \cdot 8 \cdot \ldots \cdot 8 \quad (14 \text{ factors}),$$
$$x^1 = x,$$
$$x^2 = xx,$$
$$(a+7)^3 = (a+7)(a+7)(a+7),$$
$$(xy)^3 = xyxyxy = xxxyyy = x^3 y^3.$$

It should be clear that exponents do not give rise to new expressions but furnish a new and compact way to designate old ones. The expressions in question are those which are formed by multiplications only, and in which, moreover, all the factors are the same.

We now state and prove three theorems that contain the principal facts needed for dealing with exponents.

THEOREM II: *If $E$ is any expression, and $n$ and $m$ are any nonzero cardinals, then*

$$E^n E^m = E^{n+m}.$$

*Proof*:  $E^n = EE \ldots E$         (by definition)
                ($n$ factors)

$$E^m = EE \ldots E \qquad \text{(by definition)}$$
$$(m \text{ factors})$$

$$E^n E^m = (EE \ldots E)(EE \ldots E)$$
$$(n \text{ factors, } m \text{ factors})$$

$$= EE \ldots E \ EE \ldots E \qquad \text{(GM)}$$
$$(n \text{ factors, } m \text{ factors})$$

$$= E^{n+m} \qquad \text{(definition of addition of cardinals; definition of exponents).}$$

*Example*: $2^2 2^3 = 2^{2+3} = 2^5$. This can be verified independently by computing $2^5 = 32$ and $2^2 2^3 = (4)(8) = 32$.

Theorem III: *Let $E$ be any expression, and let $n$ and $m$ be any non-zero cardinals. Then*

$$(E^n)^m = E^{nm}.$$

*Proof*: $(E^n)^m = (EE \ldots E)(EE \ldots E) \ldots (EE \ldots E)$
$$(m \text{ factors of } n \text{ factors each})$$
$$\text{(by definition of exponents)}$$

$$= EE \ldots EEE \ldots EEE \ldots E \qquad \text{(by GM)}$$

$$= E^{nm} \qquad \text{(by definition of exponents and of multiplication).}$$

*Example*: $(x^2)^3 = x^6$, which can be verified independently by writing

$$(x^2)^3 = (x^2)(x^2)(x^2) = (xx)(xx)(xx) = xxxxxx = x^6.$$

This independent verification, with appropriate justifications, actually retraces the steps of the proof.

Theorem IV: *Let $E$ and $F$ be any expressions, and let $n$ be any non-zero cardinal. Then*

$$(EF)^n = (E^n)(F^n).$$

*Proof:* $(EF)^n = (EF)(EF)\ldots(EF)$   by definition
     ($n$ factors)      of exponents)
$= EFEF\ldots EF$     (by GM)
$= EEE\ldots EFFF\ldots F$   (by GM)
$= (EE\ldots E)(FF\ldots F)$   (by GM)
$= (E^n)(F^n)$     (by definition
          of exponents).

*Example:* $(2\cdot3)^3 = (2^3)(3^3)$. This can be verified independently by computing $2\cdot3 = 6$, $6^3 = 6\cdot6\cdot6 = 216$, $2^3 = 8$, $3^3 = 27$, $8\cdot27 = 216$.

## Monomials

A *monomial* is defined to be an expression that is either a cardinal number in standard form, a single variable, or a product of cardinal numbers and variables. *Examples:* 3, 7, 7$x$, $y$6$y$, 3$xy$2$zx$. Monomials owe their importance to this simple chain of facts: Every expression has an equal expression with no parentheses; every expression with no parentheses is a sum of cardinal numbers, variables, and products of cardinal numbers and variables—that is, a sum of monomials. Thus, every expression is equal to a sum of monomials.

From a monomial that has more than one component, other monomials may be formed by rearrangement of its factors; these new monomials are, by GM, equal to the original one. We can therefore say that a monomial that depends on the variables $x, y, \ldots, w$ is equal to a monomial

$$abc\ldots gxx\ldots xyy\ldots y\ldots w\ldots w$$

where $a, b, c, \ldots, g$ are the numbers that appear in the monomial as factors, and where the rest of the symbol indicates that all the factors $x$ have been grouped together and that each of the other variables that appear as factors has been treated similarly. Such a monomial can be written in the more compact form

$$k\,x^p y^q \ldots w^r,$$

where $k$ denotes the standard form of the product $abc\ldots g$, and where $p$ is the number of times $x$ appears as a factor in the monomial, $q$ the number of times $y$ appears as a factor, and $r$ the number of times $w$ appears as a factor. We shall say that a monomial of this description is in *standard form*. We have just shown that every monomial is equal to a monomial in standard form and have shown how to find it.

*Examples*:  $x6x = 6xx = 6x^2, \quad wwx3\cdot4y2xwy$
$$= 3\cdot4\cdot2xxyywww = 24x^2y^2w^3.$$

A monomial in standard form generally has the form $aP$, where $a$ is a number (which may not appear, as in the monomial $x^2y^3$) and where $P$ is a product of powers of variables (which may not appear, as in the monomial 6). The numerical factor is called the *coefficient* of the monomial. If no coefficient appears in a monomial, we agree to say that it has coefficient 1. This agreement is based on the fact that $1E = E$, so that every monomial without a coefficient equals a monomial with coefficient 1. Another characteristic of a monomial is its *degree*, which we proceed to define. If a monomial has a coefficient that is different from zero, and if variables actually appear in it, its degree is the sum of the exponents of the variables. If the monomial is a cardinal number other than zero, its degree is defined to be 0. A monomial with coefficient zero has no degree.

*Examples*:  The degree of $3x^2y^3$ is 5; its coefficient is 3.
The degree of $3x$ is 1; its coefficient is 3.
The degree of 3 is 0; its coefficient is 3.
$0x^4y^6$ has no degree; its coefficient is 0.

All monomials with zero coefficient are equal, since $0E = 0$. We shall regard the monomial 0 as the standard form for all such monomials. The reader should note that monomials that have no coefficient are quite different from those that have the coefficient zero, even though the words suggest a similarity.

## Multiplication and Addition of Monomials

The rules that govern the multiplications and the additions of any expressions apply to those special expressions we have called monomials. Some further rules can be deduced which apply only to monomials.

The product of two monomials is evidently a monomial and is therefore equal to a monomial in standard form. It is easy to find the standard form of the product when the factors are in standard form by using the multiplication table, $GM$, and Theorem II of this chapter. For instance, the product $(3x^2y^3z^4)(4x^3z^5w^2)$ equals, by $GM$, $3 \cdot 4(x^2x^3)y^3(z^4z^5)w^2$, which equals $12x^5y^3z^9w^2$ by virtue of the multiplication table $(3 \cdot 4 = 12)$ and Theorem II $(x^2x^3 = x^5$, etc.). It follows that the coefficient of the product is the product of the coefficients of the factors and that the degree of the product is the sum of the degrees of the factors.

Addition of monomials is somewhat more complicated. It is clear that the sum of two monomials is not a monomial, because only expressions involving no additions are called monomials. However, under certain circumstances the sum of two monomials is equal to a monomial, and we proceed to investigate these circumstances.

Let $aP$ and $bQ$ be two monomials, with coefficients $a$ and $b$ respectively and products of powers of variables $P$ and $Q$. If $P$ and $Q$ are the same, we have

$$\begin{aligned} aP + bQ &= aP + bP && \text{(by hypothesis)} \\ &= (a + b)P && \text{(by CM and D),} \end{aligned}$$

which is a monomial with coefficient $(a + b)$. Such monomials, that is, two monomials with the same powers of the same variables, are called *like monomials*. Their sum is equal to a monomial.

*Examples*: $3x^2 + 5x^2 = 8x^2$.

$3x^2 + 5x$ does not equal a monomial.

## Polynomials

A *polynomial* is defined to be an expression that is either a single monomial or a sum of monomials. Theorem I allows us to conclude that for each expression there is an equal expression that is a polynomial. The proof of Theorem I also shows how to find a polynomial equal to a given expression. It follows that our problem of deciding whether or not two given expressions are equal can be reduced to the problem of deciding whether two polynomials are equal. If $E$ and $F$ are two given expressions and $P$ and $Q$ are polynomials for which $E = P$ and $F = Q$, then $E$ and $F$ are equal if and only if $P$ and $Q$ are equal We now investigate the problem of deciding when two polynomials are equal.

We begin by considering polynomials that depend on a single variable, say $x$, and define certain special polynomials of this kind. We shall say that a polynomial is in *standard form* if

1) each monomial that appears in it is in standard form;
2) no two monomials in it are like monomials;
3) its terms are arranged by degree so that of any two monomials in it the one of smaller degree appears to the left of the one of larger degree.

We now show that for every polynomial that depends on a single variable, there is an equal polynomial in standard form. For if we have a polynomial $P$ with a monomial not in standard form, we can replace this monomial by its equal in standard form, obtaining a new polynomial which, by Theorem I of Chapter III, is equal to $P$. By repeating this procedure, we can find a polynomial $Q$ that is equal to $P$ and obeys condition 1). If $Q$ has a pair of like monomials in it, we can substitute for them both the single monomial to which their sum is equal, and the new polynomial will be equal to $Q$ by GA. By repeating this procedure, we can find a polynomial $R$, equal to $Q$ (and therefore to $P$) and obeying conditions 1) and 2). If $R$ does not obey condition 3), we can rearrange its terms to obtain a polynomial $S$ which does. Such a polynomial $S$ equals $R$ by GA and hence equals $P$. The polynomial $S$ has then all the desired properties, since it equals $P$ and is in standard form.

*Example:*

$$1 + x\,3x + 2x + x^2\,5 + 6 = 1 + 3x^2 + 2x + 5x^2 + 6$$
$$= 7 + 8x^2 + 2x$$
$$= 7 + 2x + 8x^2.$$

It follows from Theorem III of Chapter III that if two polynomials are equal to the same polynomial in standard form, they are equal to each other. It is by no means so easy to show that if two polynomials are equal to different polynomials in standard form they are not equal. This is in fact the case, but we shall not give the proof until Chapter IX.

The case of polynomials that depend on more than one variable can be treated similarly. The definition of "standard form" for such polynomials contains conditions 1), 2), and 3) but requires the elaboration of item 3). For, $2x^3 y^2$ and $3 y^5$ are, for example, both monomials of degree 5 that are in standard form, but condition 3) gives no information as to which one precedes the other. We complete our discussion of this topic in Chapter IX, where we give a complete set of conditions for polynomials depending on several variables to be in standard form. We then show how any pair of expressions in several variables can be tested for equality by means of these polynomials, in a way analogous to the method used in the case of one variable.

## EXERCISES

1. For each of the following, find an equal expression having no parentheses:
   a) $x + (y + (z + w))$.
   b) $x(y + (z + w))$.
   c) $(x + y)(y + (x + w))$.
   d) $(x(y + z))(w + 3)$.
   e) $(xy(z + 1))((xyz)(xz + 1))$.

2. Find the standard form for each of the following:
   a) $xyx$.                    d) $(3x^2)(4x^3)$.
   b) $2x3x4x$.                 e) $(3x^2)^4$.
   c) $x3xyxzyxz7$.

3. Find the standard form for each of the following:
   a) $x(x + 1)$.
   b) $(2x + 3)(3x + 4)$.
   c) $(2x + 3)^2$.
   d) $(2x^2 + 3x + 4) + (x^2 + 4x + 6)$.
   e) $(2x^2 + 3x + 4) + (x^3 + 2x)$.
   f) $(2x^2 + 3x + 4)(2x + 3)$.
   g) $(2x^2 + 3x + 4)(3x^2 + 2x + 3)$.
   h) $x^2(x + 1) + x(x^2 + 1) + x^2(x^2 + 1)$.

# CHAPTER V

## NUMBER SYSTEMS IN GENERAL

### Introduction

THE PRECEDING CHAPTERS have been devoted largely to examining the familiar cardinal number system from a somewhat abstract point of view. We assume that the reader has mastered this point of view, and invite him now to apply it to a less familiar subject matter. The aim of this chapter is to describe what is understood by the terms *number* and *number system* in present-day mathematics. A century ago, this enterprise would not have been possible, because there was no agreement then on the meanings of these terms. Indeed, the names of some of the objects now accepted as numbers, such as "surds" (for absurds) and "imaginaries," reveal the antagonism their introduction evoked. The confusions have been resolved by discarding as irrelevant the terms "greater," "less," "one," "many," "unit," "magnitude," "quantity," and other similar notions often thought to be basic to the idea of number. What was retained was a coherent and precise definition. This chapter is devoted to the study of this definition and of its consequences.

### Operations on a Set

The first step in our program of achieving a definition of *number* is to define a preliminary notion, that of *operation of a set*.

DEFINITION: Let $A$ be any set. By an *operation* on $A$ we mean a function on $A \times A$ to $A$.

This amounts, roughly, to saying that an operation on a set $A$ makes a member of $A$ correspond to each ordered pair composed of members of $A$. There are two ready-made examples of operations on the set of all cardinal numbers, namely the addition and the multiplication operations. It is easy to see that addition of cardinal numbers satisfies our definition of operation on a set, for addition assigns to each ordered pair of cardinal numbers $(a, b)$ the cardinal number $a + b$. Multiplication of cardinal numbers, in assigning to each ordered pair $(a, b)$ of cardinals, the cardinal $ab$, likewise defines an operation on the set of all cardinal numbers. Other familiar examples will perhaps occur to the reader.

There is nothing in the definition that requires that the set on which the operation is defined should have anything to do with numbers. Any set whatever can have operations defined on it. Furthermore there need be no discernible system in making up the assignments. A simple example can be constructed using a set with two members, say $a$ and $b$. Let us call this set $S$; to construct an operation on $S$ means to assign to each member of $S \times S$ a member of $S$. The set $S \times S$ consists of the ordered pairs $(a, a)$, $(a, b)$, $(b, a)$, $(b, b)$, and by assigning, arbitrarily, $a$ to $(a, a)$, $a$ to $(a, b)$, $b$ to $(b, a)$, and $a$ to $(b, b)$, we have constructed an operation on $S$. There are in fact fifteen other operations on $S$. Just as the operation of addition is indicated by the sign $+$, it is sometimes convenient to designate an arbitrary operation on an arbitrary set by some symbol. The operation on $S$ just described might, for instance, be named the $\#$-operation, and then we could write

$$a \# a = a,$$
$$a \# b = a,$$
$$b \# a = b,$$
$$b \# b = a.$$

Going one step further in the analogy with addition, we could display this operation by a table

| $\#$ | $a$ | $b$ |
|---|---|---|
| $a$ | $a$ | $a$ |
| $b$ | $b$ | $a$. |

## Commutative and Associative Operations

Let $S$ be any set, and let * be any operation on $S$. We shall say that * is *commutative* if

$$x * y = y * x,$$

where $x$ and $y$ are variables each of whose domains is the set $S$. Here, as usual, the sign $=$ means "the same as."

We shall say that * is *associative* if

$$x * (y * z) = (x * y) * z,$$

where $x$, $y$, and $z$ are variables each of whose domains is the set $S$. In this last definition, the parentheses are used to indicate that what they enclose is a member of $S$. Thus, $x * (y * z)$ means the member of $S$ assigned by * to the ordered pair $(x, y * z)$ of $S \times S$.

Let us examine the operation given by the table at the end of the last paragraph to determine if it is commutative or associative. It is clearly not commutative, because $a \# b = a$, $b \# a = b$, so that $a \# b$ and $b \# a$ are not the same. Neither is it associative, because

$$b \# (a \# b) = b \# a \quad \text{(because } a \# b = a\text{)}$$
$$= b,$$

whereas

$$(b \# a) \# b = b \# b \quad \text{(because } b \# a = b\text{)}$$
$$= a,$$

so that $b \# (a \# b)$ and $(b \# a) \# b$ are not the same.

For an example of an operation that is both commutative and associative, let us use the same set $S$ and a different operation $O$, given by the table

| O | a | b |
|---|---|---|
| a | a | b |
| b | b | a. |

To show that this operation is commutative, we must verify each of the four statements abbreviated by $x \, O \, y = y \, O \, x$; and to show that it is associative, we must verify each of the eight statements abbreviated by $x \mathrel{O} (y \mathrel{O} z) = (x \mathrel{O} y) \mathrel{O} z$. We shall test one of each, leaving the remaining details as an exercise for the reader.

That $a \mathrel{O} b = b \mathrel{O} a$ follows from the table; they both equal $b$. That $a \mathrel{O} (b \mathrel{O} b) = (a \mathrel{O} b) \mathrel{O} b$ follows from the equality of the first to $a \mathrel{O} a$, which is $a$, and from the equality of the second to $b \mathrel{O} b$, which is $a$. It is also possible to find operations that are associative but not commutative and operations that are commutative but not associative.

## Distributive Operations

Given a set $S$ with two operations, say * and $O$, we shall say that * is distributive relative to $O$ if

$$x * (y \mathrel{O} z) = (x * y) \mathrel{O} (x * z),$$

where $x$, $y$, and $z$ are variables each of whose domains is $S$. Here again, the sign $=$ means "the same as," and parentheses enclose single members of $S$. The reader is familiar with an example of distributive operations. If we take for $S$ the set of cardinal numbers, for * the multiplication operation, and for $O$ the addition operation, then the * thus defined is distributive relative to the $O$ thus defined. Note that in this instance $O$ is not distributive relative to *, for the condition $x \mathrel{O} (y * z) = (x \mathrel{O} y) * (x \mathrel{O} z)$ would require such absurdities as $2 + (3 \cdot 5) = (2 + 3) \cdot (2 + 5)$. A less familiar example of distributive operations can be found by combining the operation $O$ whose table is given above and the operation *, on the same set, whose table is

| * | a | b |
|---|---|---|
| a | a | a |
| b | a | b. |

To verify that this * is distributive relative to this $O$ we would have to verify eight relations, a typical one of which is

$$a * (b \mathrel{O} a) = (a * b) \mathrel{O} (a * a).$$

Let us examine this one only.

$$a * (b \bigcirc a) = a * b \qquad\qquad (b \bigcirc a = b)$$
$$= a$$

and

$$(a * b) \bigcirc (a * a) = (a * b) \bigcirc a \qquad (a * a = a)$$
$$= a \bigcirc a \qquad\qquad (a * b = a)$$
$$= a,$$

whence

$$a * (b \bigcirc a) = (a * b) \bigcirc (a * a).$$

## Numbers and Number Systems

We are now able to give the promised definitions of number and number systems. We shall give first our definition of a number system and then define a number simply as a member of a number system.

DEFINITION: Let $S$ be a set and let $*$ and $\bigcirc$ be operations on $S$. If $*$ and $\bigcirc$ are each both commutative and associative and if $*$ is distributive relative to $\bigcirc$, we say that $S$ is a *number system,* that $*$ is its *multiplication operation,* and that $\bigcirc$ is its *addition operation.*

The cardinals constitute a number system in this sense. What is perhaps more striking is that the set $S$ consisting of two members $a$ and $b$, with operations $*$ and $\bigcirc$ as given in the last section, also is a number system. There are infinitely many different number systems, in this sense. It will be our goal to construct some of the more useful ones and examine their properties.

## Zero-Element and One-Element of a Number System

The reader may have observed that our test for a set to be a number system is to see if it behaves like the cardinal number system relative to the five fundamental laws. In the same spirit, we shall define a "zero element" of a number system, and a "one element" to be numbers that behave in certain respects like the cardinals zero and one. The cardinal number zero is the cardinal number of the empty set, and thus bears a definition compatible with its meaning in everyday language. However, to obtain a definition applicable to all number systems, we abandon the everyday usage and give a definition that is based on their addition tables.

Definition : Let $N$ be a number system. A number $z$ of $N$ will be called a *zero element* of $N$ if for each member $a$ of $N$

$$z + a = a + z = a.$$

Note that the system of cardinal numbers contains a number that meets these requirements, namely the cardinal number zero. This definition states, roughly, that a given number of a given number system is the zero element of that number system if its behavior in that system is the same as the behavior of the cardinal zero in the system of cardinals. Our definition of the cardinal number one was also in accord with the meaning given the term in everyday language. Here too, in undertaking to define the notion of a number "one" for any number system, we abandon the everyday meaning and adopt a behavioristic one.

Definition : Let $N$ be a number system. A number $z$ of $N$ will be called a *one element* of the number system $N$ if for every number $a$ of $N$,

$$z \cdot a = a \cdot z = a.$$

Note that the system of cardinals has such a number, namely the cardinal number one.

The obvious similarity between the two definitions just given can be described precisely in terms of still another object.

Definition : Let $N$ be a set and let $O$ be an operation on $N$. A member $z$ of $N$ will be called an *identity element* of $N$ relative to $O$ if for each member $a$ of $N$,

$$a \, O \, z = z \, O \, a = a.$$

Our definitions of zero element and one element thus state that each is the identity element of $N$ relative to the appropriate operation.

If $N$ is a set and O is an operation on $N$, there may be no identity element of $N$ relative to O. This is the case, for instance, in the following operation:

| O | $p$ | $q$ |
|---|---|---|
| $p$ | $q$ | $q$ |
| $q$ | $q$ | $p$ |

However, the next theorem shows that if there is such an identity element it is unique.

THEOREM: *Let $N$ be a set, let O be an operation on $N$, and let $z$ and $z'$ be identity elements of $N$ relative to O. Then $z = z'$.*

*Proof:* Because $z$ is an identity element, we have $z \text{ O } z' = z'$. Because $z'$ is an identity element, we have $z \text{ O } z' = z$. It follows that $z = z'$.

COROLLARY: *A number system has at most one zero-element.*

COROLLARY: *A number system has at most one one-element.*

## Inverses, Negatives, and Reciprocals

Subtraction and division have not yet been discussed, either for cardinals or for number systems in general. Our treatment of these two activities is based on a concept belonging to the study of operations in general, namely the concept of the *inverse*.

DEFINITION: Let $N$ be a set, let * be an operation on $N$, let $z$ be the identity element of $N$ relative to *, and let $a$ be a member of $N$. Then if there is a member $b$ of $N$ such that $a * b = b * a = z$, we say that $b$ is an *inverse* of $a$, and that $a$ is an *inverse* of $b$, relative to *.

THEOREM: *If * is an associative operation on a set $N$, then each member of $N$ has at most one inverse, relative to *.*

*Proof:* Let $z$ be the identity element of $N$, relative to *, let $a$ be any member of $N$, and let $b$ and $c$ be inverses of $a$. Then

$$b = z * b \qquad (z \text{ is the identity})$$
$$= (c * a) * b \qquad (c \text{ is an inverse of } a)$$
$$= c * (a * b) \qquad (* \text{ is associative})$$
$$= c * z \qquad (b \text{ is an inverse of } a)$$
$$= c \qquad (z \text{ is the identity})$$

It is in terms of inverses that we define negatives and reciprocals.

DEFINITION: Let $N$ be a number system with a zero-element. Then by the *negative* of a number of $N$ we mean its inverse relative to the addition operation of $N$.

DEFINITION: Let $N$ be a number system with a one-element. Then by the *reciprocal* of a number of $N$ we mean its inverse relative to the multiplication operation of $N$.

The concept of *negative* was not used in our study of the cardinal number system because there is only one cardinal number which has a negative, namely zero. This concept will be important in our study of other number systems. Beginners in algebra are often asked to identify the notion of the negative of a number with some situation involving a deficiency, such as overdrawn bank balances or temperatures below zero. We urge the student not to do this, and to regard the rather colorless definition given above as providing a complete and accurate version of this concept.

## Some Remarks About Numbers in General

Most of the remainder of this book is given over to the construction and study of certain individual number systems, namely, the integers, the rationals, and the reals. Each of these number systems plays a useful role in many branches of contemporary life and is successfully employed in many instances by people who have no acquaintance with the points of view we employ here. The reader whose sole concern with numbers and with mathematics lies in their application to human affairs can find briefer routes to his goal than the following chapters. What we shall do for each of the number systems mentioned is to build it up by constructing a suitable set and defining suitable operations on the set. We shall then derive by logical inference, rather than learn by rote, the special properties of each of the number systems that we shall construct. The reader who follows us in this exposition can not only learn how to use numbers but will be exposed to a genuine and typical instance of a mathematical problem and a mathematical procedure.

## EXERCISES

1. a) List all operations on the set $\{a, b\}$   (16 answers).
    b) Determine which of these are commutative.
    c)· Find one which is associative and one which is not associative.
    d) Determine which of these have identity elements.
    e) Determine which of those found in d) have an inverse for each member.
    f) Find one which has a cancellation law and one which does not have a cancellation law.

2. Let $S$ be a set and let $T$ be the set of all subsets of $S$. Assign to each pair of subsets of $S$ the union of that pair. Show that this defines an operation on $T$ that is commutative and associative and has an identity element.

3. Do the same as 2. for intersections.

4. Show that the intersection operation (see Ex. 3.) is distributive relative to the union operation and also that the union operation (see Ex. 2.) is distributive relative to the intersection operation.

5. Prove that the cardinal number system is a number system, that the cardinal 0 is its zero-element, and that the cardinal 1 is its one-element.

6. a) Prove that the set $\{a, b\}$ with operations

| $O$ | $a$ $b$ |
|---|---|
| $a$ | $a$ $a$ |
| $b$ | $a$ $b$ |

| $*$ | $a$ $b$ |
|---|---|
| $a$ | $a$ $b$ |
| $b$ | $b$ $a$ |

is a number system, with zero-element $a$ and one-element $b$.

b) Prove also

that $a$ is the negative of $a$,
that $b$ is the negative of $b$,
that $b$ is the reciprocal of $b$.
that $a$ has no reciprocal.

7. a) Construct a number system using the results of 2, 3, 4 above.

b) Find the zero-element and the one-element for the number system constructed in a). Observe that most members have neither a negative nor a reciprocal.

c) Consider the special case where $S$ is a set whose only member is $p$. Construct tables for the union and intersection operations in this case.

8. Prove that a number system can have at most one zero-element and at most one one-element.

9. Let $S$ be the set $\{0, 2, 4, 6, \cdots\}$. Let * and O be defined as ordinary addition and multiplication. Prove that, relative to these operations,

a) $S$ is a number system;
b) $S$ has a zero-element;
c) $S$ has no one-element.

10. Let $C$ be the set of all non-zero cardinal numbers. We now define the exponential operation on $C$. If $a$ and $b$ are any members of $C$, to the pair $(a, b)$ assign the non-zero cardinal number $a^b$.

a) Show by an example that this operation is not commutative and not associative.

b) Is 1 an identity element for this operation?

11. Let $S$ be a set and let * be an operation on $S$ such that:

i) * is associative;
ii) * has an identity element $e$;
iii) Each member of $S$ has an inverse element.

Prove that this operation has a cancellation law. (In other words, prove that if $a * b = a * c$, then $b = c$.)

12. Let $C$ be the set of all cardinal numbers. We consider subsets $S$ of $C$ having the property that the ordinary addition operation on $C$ defines an operation on $S$. (An example of such a set is given in Exercise 9 of Chapter V.)

a) Find other examples of such sets.

b) If $S$ and $T$ are sets with this property, does $S \cup T$ have the property? Does $S \cap T$ have the property?

# CHAPTER VI

## CONSTRUCTION OF THE INTEGERS

### Introduction

THIS CHAPTER IS GIVEN OVER TO the construction of an important
number system called the integers. We assemble a set of objects,
define two operations on this set, and show that the five fundamental
laws are obeyed. Each step in this program has many details, and
the reader may find it helpful to keep the overall program in mind as
he deals with them.

When we compare the new number system with the cardinals
we discover that the integers constitute a kind of enlargement of
the system of cardinals. A great many facts about cardinals can be
carried over to the new system, and the new system has useful new
properties of its own. When we constructed the cardinals we were
never very far removed from common sense and everyday experience. The terminology used was perhaps novel, but the concepts
and procedures were, in essentials, borrowed from familiar situations. Our work with the integers is far more abstract and far less
transparent. The reader who finds our definitions arbitrary and
somewhat bizarre should be informed that they are the product of
a long evolution of which our exposition gives only the final stage.

### Definition of Integers

We use the set of all ordered pairs of cardinal numbers to obtain
the integers. The set of all ordered pairs of cardinal numbers has
an endless membership list: $(0,0)$, $(0,1)$, $(0,2)$, ..., $(1,0)$,

$(1, 1)$, ..., $(15, 87)$, ...; we shall select certain subsets of this set and call them integers.

DEFINITION: For each ordered pair of cardinal numbers $(a, b)$, we designate by $I(a, b)$ the set of all ordered pairs of cardinals $(u, v)$ for which

$$a + v = b + u.$$

Each such set $I(a, b)$ will be called an *integer*.

*Example*: $I(7, 11)$ has the ordered pair $(8, 12)$ as a member, because $7 + 12 = 11 + 8$. The ordered pair $(5, 2)$ is not a member of $I(7, 11)$, because $7 + 2$ and $11 + 5$ are not equal. There are countless other ordered pairs that are members of $I(7, 11)$, and countless other ordered pairs that are not members of $I(7, 11)$. Let us designate the set of all cardinal numbers by $C$; then $C \times C$ designates the set of all ordered pairs of cardinals. Each integer is a certain subset of the set $C \times C$, and for each member $(a, b)$ of $C \times C$ there is an integer $I(a, b)$.

## Relations Among the Sets $I(x, y)$

It is possible for two integers with different designations, like $I(5, 6)$ and $I(8, 9)$ to be the same. Both $I(5, 6)$ and $I(8, 9)$ are sets, and we have agreed that any two sets are the same if they have the same members. To see that $I(5, 6)$ and $I(8, 9)$ are the same, note that if $(u, v)$ is a member of $I(5, 6)$, then

$$5 + v = 6 + u \qquad \text{(by definition of } I(5, 6)\text{)}$$
$$3 + 5 + v = 3 + 6 + u$$
$$8 + v = 9 + u,$$

and therefore $(u, v)$ is a member of $I(8, 9)$. On the other hand, if $(r, s)$ is a member of $I(8, 9)$, then

$$8 + s = 9 + r \qquad \text{(by definition of } I(8, 9)\text{)}$$
$$3 + 5 + s = 3 + 6 + r$$
$$5 + s = 6 + r \qquad \text{(cancellation law for addition).}$$

Therefore $(r, s)$ belongs to $I(5, 6)$. We have shown that every member of $I(5, 6)$ is a member of $I(8, 9)$ and that every member of $I(8,9)$ is a member of $I(5,6)$. This proves that $I(5,6)$ and $I(8,9)$ are the same set.

It is possible for two integers to have no ordered pairs in common. For instance, $I(5,6)$ and $I(5,7)$ have no members in common. To prove this, let $(u,v)$ be a member of $I(5,6)$. Then

$$5 + v = 6 + u.$$

We could not also have

$$5 + v = 7 + u,$$

which would be required for $(u,v)$ to be a member of $I(5,7)$. We now show that these two examples exhibit the only possible relations among the sets $I(x,y)$.

THEOREM I: *If $I(a,b)$ and $I(c,d)$ are integers, then either $I(a,b)$ and $I(c,d)$ are the same, or $I(a,b)$ and $I(c,d)$ have no members in common.*

*Note*: The point of this theorem is that two integers cannot have in common some, but not all, of their ordered pairs.

*Proof*: Assume the second alternative does not hold; that is, assume there is an ordered pair $(p,q)$ which is a member both of $I(a,b)$ and of $I(c,d)$. We show that the first alternative then must hold: that $I(a,b)$ and $I(c,d)$ are the same. We have

$$a + q = b + p \quad ((p,q) \text{ is a member of } I(a,b)),$$
$$c + q = d + p \quad ((p,q) \text{ is a member of } I(c,d)).$$

To show that $I(a,b)$ and $I(c,d)$ are the same, it is sufficient to show that each member of $I(a,b)$ is a member of $I(c,d)$. Let $(r,s)$ be any member of $I(a,b)$. We are to show that $(r,s)$ is also a member of $I(c,d)$.

$$a + s = b + r \quad ((r,s) \text{ is a member of } I(a,b)).$$

Then

$$(b+p) + (c+q) + (a+s) = (a+q) + (d+p) + (b+r)$$

and

$$(c+s) + (a+b+p+q) = (d+r) + (a+b+p+q)$$
$$\text{(by GA)}$$

whence

$$c + s = d + r \qquad \text{(by cancellation law}$$
$$\text{for addition).}$$

This proves that $(r, s)$ is in $I(c, d)$.

We supplement this theorem with the observation that every ordered pair $(a, b)$ is in some set $I(x, y)$, namely, $I(a, b)$. To prove this, apply the test for the membership of $(a, b)$ in $I(a, b)$; this requires $a + b = b + a$, which is guaranteed by CA. Thus, every member of $C \times C$ is in one and only one set $I(x, y)$. The set $I(x, y)$ to which a member of $C \times C$ belongs may have many different designations, but any two sets $I(a, b)$ and $I(c, d)$ that contain a given member of $C \times C$ must be the same, by Theorem I.

It is useful to have a test to decide whether two sets $I(a, b)$ and $I(c, d)$ are the same or not. The following theorem provides such a test.

THEOREM II: $I(a, b)$ and $I(c, d)$ are the same if $a + d = b + c$; $I(a, b)$ and $I(c, d)$ are not the same (and hence have no members in common) if $a + d$ is not equal to $b + c$.

*Proof*: We know that the ordered pair $(c, d)$ is a member of $I(c, d)$. By definition, the ordered pair $(c, d)$ is a member of $I(a, b)$ if $a + d = b + c$, and is not a member of $I(a, b)$ if $a + d$ is not equal to $b + c$. The first alternative implies that $(c, d)$ is a common member of the sets $I(a, b)$ and $I(c, d)$, which, by Theorem I, implies that $I(a, b)$ and $I(c, d)$ are the same. The second alternative implies that $I(a, b)$ and $I(c, d)$ are not the same, since $(c, d)$ is a member of one and not of the other. This proves the theorem.

## Definition of Addition for Integers

We now define an operation on the set of integers.

DEFINITION: Let $I(a, b)$ and $I(c, d)$ be any integers. We define as their "*sum*" and designate by the symbol $I(a, b)$ "$+$" $I(c, d)$, the integer $I(a + c, b + d)$.

*Example:* Given the specific integers $I(3, 7)$ and $I(6, 12)$, their "sum" $I(3, 7)$ "$+$" $I(6, 12)$ is the integer $I(9, 19)$.

We have used quotation marks around the word "sum" and the symbol " + " to distinguish these items from the word sum and the symbol + as employed with cardinals. An inspection of the definition of "sum" and " + " will show that these are actually defined in terms of sum and + ; hence the need for the distinction.

Our definition tells us how to obtain, from any two symbols that designate integers, a new symbol that designates an integer. We now raise a question about this definition which is based on the fact that different symbols may designate the same integer. For example, consider

$$I(5,6) \text{ " + " } I(1,4) = I(6,10)$$

and

$$I(8,9) \text{ " + " } I(1,4) = I(9,13)$$

In both "sums" the integers on the left are the same, since $I(5,6)$ and $I(8,9)$ have been shown to be the same. Thus we have found the "sum" of a single pair of integers in two different ways. Our question is whether or not the integers obtained in these two ways are the same, that is, whether $I(6,10)$ is the same as $I(9,13)$. The question is important, because we have claimed to be dealing with integers, and the "sum" of two integers ought therefore depend only on these integers and not on the particular symbols used to designate them. In other words, to justify our definition of "sum" as a definition about integers rather than about symbols for integers, we must prove the following theorem.

THEOREM III: *If* $I(a,b) = I(c,d)$ *and if* $I(p,q) = I(r,s)$, *then*

$$I(a,b) \text{ " + " } I(p,q) = I(c,d) \text{ " + " } I(r,s).$$

*Proof:* By definition

$$I(a,b) \text{ " + " } I(p,q) = I(a+p,b+q)$$
$$I(c,d) \text{ " + " } I(r,s) = I(c+r,d+s).$$

What we are to show is

(1) $$I(a+p,b+q) = I(c+r,d+s).$$

We see from Theorem II that in order to prove (1), it suffices to show that

(2)         $(a + p) + (d + s) = (b + q) + (c + r).$

To prove (2), observe that from $I(a, b) = I(c, d)$ and Theorem II, we have

(3)                    $a + d = b + c,$

and similarly from $I(p, q) = I(r, s)$ we have

(4)                    $p + s = q + r.$

Combining (3) and (4), we have

(5)         $(a + d) + (p + s) = (b + c) + (q + r),$

and (2) follows by applying GA to (5).

## The Commutative and Associative Laws for Addition of Integers

We now state and prove two theorems about "sums" of integers, whose relation to CA and AA for cardinals should be obvious.

THEOREM IV:
$$X \text{ "+" } Y = Y \text{ "+" } X,$$

*where $X$ and $Y$ are variables each of whose domains is the set of integers.*

*Note:* This theorem is an abbreviation for an endless list of statements

$$I(0, 0) \text{ "+" } I(0, 1) = I(0,1) \text{ "+" } I(0, 0)$$
. . . . . . . . . . . . . . . . . . . . . . . . . . . . . . . . . . .
. . . . . . . . . . . . . . . . . . . . . . . . . . . . . . . . . . .
. . . . . . . . . . . . . . . . . . . . . . . . . . . . . . . . . . . . . .

$$I(35, 68) \text{ "+" } I(19, 42) = I(19, 42) \text{ "+" } I(35, 68)$$
. . . . . . . . . . . . . . . . . . . . . . . . . . . . . . . . . . . .

We reformulate the theorem in terms of other variables.

Theorem IV (restated):

$$I(x, y) \text{ "}+\text{" } I(p, q) = I(p, q) \text{ "}+\text{" } I(x, y),$$

*where $x$, $y$, $p$, $q$ are variables each of whose domains is the set of cardinal numbers.*

*Proof:*

$$I(x, y) \text{ "}+\text{" } I(p, q) = I(x + p, y + q) \qquad \text{(by definition)}$$
$$I(p, q) \text{ "}+\text{" } I(x, y) = I(p + x, q + y) \qquad \text{(by definition)}.$$

To prove these equal we use Theorem II, deducing from this theorem that $I(x + p, y + q) = I(p + x, q + y)$ if $(x + p) + (q + y) = (p + x) + (y + q)$. But this follows from GA, and thus our theorem is proved. We shall refer to this theorem as law CA for integers.

Theorem V:

$$X \text{ "}+\text{" } (Y \text{ "}+\text{" } Z) = (X \text{ "}+\text{" } Y) \text{ "}+\text{" } Z,$$

*where $X$, $Y$, and $Z$ are variables each of whose domains is the set of integers.*

*Note:* This theorem is an abbreviation for an endless list of statements, a typical one of which is

$$I(2, 6) \text{ "}+\text{" } (I(5, 1) \text{ "}+\text{" } I(7, 3))$$
$$= (I(2, 6) \text{ "}+\text{" } I(5, 1)) \text{ "}+\text{" } I(7, 3).$$

We reformulate the theorem in terms of other variables.

Theorem V (restated):

$$I(x, y) \text{ "}+\text{" } (I(p, q) \text{ "}+\text{" } I(z, w))$$
$$= (I(x, y) \text{ "}+\text{" } I(p, q)) \text{ "}+\text{" } I(z, w),$$

*where $x$, $y$, $p$, $q$, $z$, $w$ are variables each of whose domains is the set of cardinal numbers.*

*Proof:* $\quad I(p, q) \text{ "}+\text{" } I(z, w) = I(p + z, q + w)$
$$\text{(by definition)}$$

$$I(x, y) \text{ "}+\text{" } I(p + z, q + w) = I(x + (p + z), y + (q + w)).$$

Similarly, we find that

$$(I(x, y) \text{ "} + \text{" } I(p, q)) \text{ "} + \text{" } I(z, w)$$
$$= I((x + p) + z, (y + q) + w).$$

To prove the theorem, we need to show that

$$I(x + (p + z), y + (q + w)) = I((x + p) + z, (y + q) + w).$$

To establish this, we need only show (using Theorem II) that

$$x + (p + z) + (y + q) + w = y + (q + w) + (x + p) + z,$$

and this follows from GA. We shall refer to this theorem as law AA for integers.

### Definition of Multiplication for Integers

We can now introduce our second operation on the set of integers.

DEFINITION: Let $I(a, b)$ and $I(c, d)$ be any integers. We define as their "*product*" and designate by the symbol $I(a, b) \text{ "} \cdot \text{" } I(c, d)$, the integer $I(ac + bd, ad + bc)$.

*Example*: For the specific integers $I(3, 7)$ and $I(6, 12)$, this definition implies that their "product" $I(3, 7) \text{ "} \cdot \text{" } I(6, 12)$ is the integer $I(102, 78)$.

As before, quotation marks distinguish operations involving integers from those involving cardinals. This definition creates a problem very much like one we met in dealing with addition of integers. Our definition of "product" claims to be a definition for integers, and yet it is based on the particular symbols used to designate the integers. Since the same integers can be designated by many different symbols, the question arises as to whether or not the "product" of two integers is independent of the particular symbols used to designate them. The following theorem formulates this question in a more precise way and answers it affirmatively.

THEOREM VI: *If $I(a, b) = I(c, d)$ and if $I(p, q) = I(r, s)$, then*

$$I(a, b) \text{ "} \cdot \text{" } I(p, q) = I(c, d) \text{ "} \cdot \text{" } I(r, s).$$

*Proof:* First, we show that

(1)  $I(a, b) \text{``·''} I(p, q) = I(c, d) \text{``·''} I(p, q).$

We have

$. I(a, b) \text{``·''} I(p, q) = I(ap + bq, aq + bp)$  (by definition)

$I(c, d) \text{``·''} I(p, q) = I(cp + dq, cq + dp)$  (by definition)

(2)  $a + d = b + c$  (Theorem II applied to $I(a, b) = I(c, d)$).

To establish (1), we write

$$ap + bq + cq + dp = (a + d)p + (b + c)q \quad \text{(by D and GA)}$$
$$= (b + c)p + (a + d)q \quad \text{(by 2)}$$
$$= aq + bp + cp + dq \quad \text{(by D and GA)}.$$

Thus,

$$ap + bq + cq + dp = aq + bp + cp + dq$$

which, by Theorem II, proves that

$$I(ap + bq, aq + bp) = I(cp + dq, cq + dp).$$

Similarly we can show that

$$I(c, d) \text{``·''} I(p, q) = I(c, d) \text{``·''} I(r, s),$$

which, with (1), implies what we wish to prove.

## The Commutative and Associative Laws for Multiplication of Integers

We now take up the analogues for integers of laws CM and AM.

THEOREM VII:
$$X \text{``·''} Y = Y \text{``·''} X,$$

*where X and Y are variables each of whose domains is the set of integers.*

*Note:* This theorem is an abbreviation for an endless list of statements, a typical one of which is

$$I(35, 68) \text{``·''} I(19, 42) = I(19, 42) \text{``·''} I(35, 68).$$

We reformulate the theorem in terms of other variables.

Theorem VII′:

$$I(x, y) \text{ “•” } I(p, q) = I(p, q) \text{ “•” } I(x, y),$$

*where x, y, p, q are variables each of whose domains is the set of all cardinal numbers.*

*Proof:*    $I(x, y) \text{ “•” } I(p, q) = I(xp + yq, xq + yp).$
           $I(p, q) \text{ “•” } I(x, y) = I(px + qy, py + qx)$
                                     (by definition).

To prove these equal, we use Theorem II, deducing from it that

$$I(xp + yq, xq + yp) = I(px + qy, py + qx)$$

if

$$(xp + yq) + (py + qx) = (xq + yp) + (px + qy).$$

But this follows from CM and GA, and thus our theorem is proved. We shall refer to this theorem as law CM for integers.

Theorem VIII:

$$X \text{ “•” } (Y \text{ “•” } Z) = (X \text{ “•” } Y) \text{ “•” } Z,$$

*where X, Y, and Z are variables whose domain is the set of integers.*

*Note:*   This theorem is an abbreviation for an endless list of statements, a typical one of which is

$$I(2, 6) \text{ “•” } (I(5, 1) \text{ “•” } I(7, 3))$$
$$= (I(2, 6) \text{ “•” } I(5, 1)) \text{ “•” } I(7, 3).$$

We reformulate the theorem in terms of other variables.

Theorem VIII′: *Let x, y, p, q, r, s be variables each of whose domains is the set of cardinal numbers. Then*

$$I(x, y) \text{ “•” } (I(p, q) \text{ “•” } I(r, s))$$
$$= (I(x, y) \text{ “•” } I(p, q)) \text{ “•” } I(r, s).$$

*Proof:*    $I(p, q) \text{ “•” } I(r, s) = I(pr + qs, ps + qr)$
                             (by definition of "multiplication")

(1)    $I(x, y)$ "$\cdot$" $(I(p, q)$ "$\cdot$" $I(r, s))$
     $= I(x(pr + qs) + y(ps + qr), x(ps + qr) + y(pr + qs))$
                        (by definition of "multiplication").

Similarly we find that

(2)    $(I(x, y)$ "$\cdot$" $I(p, q))$ "$\cdot$" $I(r, s)$
     $= I((xp + yq)r + (xq + yp)s, (xp + yq)s + (xq + yp)r)$

The right member of (1) is found, by D, to equal

(3)        $I(xpr + xqs + yps + yqr, xps + xqr + ypr + yqs)$

and the right member of (2) is found to equal

(4)        $I(xpr + yqr + xqs + yps, xps + yqs + xqr + ypr)$

and (3) and (4) are the same, by GA. This proves the theorem. We shall refer to this theorem as law A M for integers.

## The Distributive Law for Integers

The analogue for integers of Law D is

Theorem IX:

$$X \text{ "}\cdot\text{" } (Y \text{ "}+\text{" } Z) = (X \text{ "}\cdot\text{" } Y) \text{ "}+\text{" } (X \text{ "}\cdot\text{" } Z),$$

*where X, Y, and Z are variables each of whose domains is the set of integers.*

We reformulate the theorem in terms of new variables.

Theorem IX′: *Let x, y, p, q, r, s be variables each of whose domains is the set of cardinal numbers. Then*

   $I(x, y)$ "$\cdot$" $(I(p, q)$ "$+$" $I(r, s))$
     $= (I(x, y)$ "$\cdot$" $I(p, q))$ "$+$" $(I(x, y)$ "$\cdot$" $I(r, s))$.

*Proof:*     $I(p, q)$ "$+$" $I(r, s) = I(p + r, q + s)$
                   (by definition of "addition" for integers)

   $I(x, y)$ "$\cdot$" $I(p + r, q + s)$
     $= I(x(p + r) + y(q + s), x(q + s) + y(p + r))$
                   (by definition of "multiplication" for integers)

whence, by D,

(1)    $I(x, y) \text{ "} \cdot \text{" } (I(p, q) + I(r, s))$
$$= I(xp + xr + yq + ys, xq + xs + yp + yr).$$

On the other hand, we have

$$I(x, y) \text{ "} \cdot \text{" } I(p, q) = I(xp + yq, xq + yp)$$
(by definition of "multiplication"
for integers)

$$I(x, y) \text{ "} \cdot \text{" } I(r, s) = I(xr + ys, xs + yr)$$
(by definition of "multiplication"
for integers)

(2)    $(I(x, y) \text{ "} \cdot \text{" } I(p, q)) \text{ "} + \text{" } (I(x, y) \text{ "} \cdot \text{" } I(r, s))$
$$= I(xp + yq + xr + ys, xq + yp + xs + yr).$$

But by GA, the right members of (1) and (2) are equal. This completes the proof.

We shall refer to this theorem as law D for integers.

### Relation Between Cardinals and Integers; Isomorphisms

We have shown that our set of integers, with the two operations "$+$" and "$\cdot$", forms a number system. There remains the task of showing what claim this particular number system has on our attention. We shall now show that the integers are, at the very least, as useful as the cardinals; we accomplish this by showing that the integers can be made to do whatever the cardinals do. Our next chapter shows how they can be made to do what the cardinals fail to do. We are going to exhibit an intimate relation between the cardinals and the integers. To this end, we consider a certain subset of the integers, namely, those special integers of the form $I(a, 0)$, where $a$ is any cardinal. These special integers are an endless set $I(0, 0)$, $I(1, 0)$, $I(2, 0)$, $I(3, 0)$, $\cdots$. It is clear that a one-to-one correspondence is set up between the set of all cardinals and this set of integers by the rule that the cardinal $a$ is to be paired with the integer $I(a, 0)$. It is instructive to "add" and to "multiply" these special integers. We have

$$I(a,0) \text{ "}+\text{" } I(b,0) = I(a+b, 0+0) \quad \text{(by defini-}$$
$$\text{tion of "addition")}$$

$$= I(a+b, 0) \quad (0+0=0)$$

and

$$I(a,0) \text{ "}\cdot\text{" } I(b,0) = I(ab+00, a\cdot0+b\cdot0)$$
$$\text{(by definition of "multi-}$$
$$\text{plication" for integers)}$$

$$= I(ab, 0) \quad \text{(properties of 0)}.$$

Note first that the "sum" of special integers is a special integer, and that the "product" of special integers is likewise a special integer. Note also the close connection between addition of cardinals and "addition" of corresponding integers, and the close connection between multiplication of cardinals and "multiplication" of corresponding integers. If $a$ and $b$ are any cardinals, the integer $I(a+b, 0)$ which corresponds to their sum is the same integer as the "sum" of the integers $I(a,0)$ and $I(b,0)$ which correspond to $a$ and $b$. Similarly, the integer $I(ab, 0)$ which corresponds to the product $ab$ is the same as the integer which is the "product" of $I(a,0)$ and $I(b,0)$. We shall describe this relation between the system of cardinals and the system of special integers by saying that these two systems are *isomorphic*, and that the one-to-one correspondence between the two number systems is an *isomorphism*. One of the consequences of this relation is that the addition and multiplication tables for these two sets are essentially the same.

It is in this sense that we can say that the system of integers is a kind of enlargement of the system of cardinals. On the one hand, the integers contain the subset of special integers, which is indistinguishable in action from the cardinals. There are, moreover, integers which are not special integers, that is, which are not equal to integers of the form $I(a,0)$. Thus the integers are actually a larger system than the cardinals. In Chapter VII we examine the whole system of integers in greater detail. The integers that are not special integers receive due attention there.

We now give a formal definition of isomorphism as a possible relation between any two number systems.

DEFINITION: Let $M$ and $N$ be number systems and let there be given a one-to-one correspondence between them· Call the member of $N$ that corresponds to each member of $M$ the *image* of that member. Then the given one-to-one correspondence is an *isomorphism,* and the number systems are *isomorphic,* if

(a) the image of the sum of each pair of members of $M$ is the sum of their images,

(b) the image of the product of each pair of members of $M$ is the product of their images.

Isomorphic number systems are often considered, from the algebraic point of view, as being the same. Possible similarities and differences between such systems are well illustrated by comparing the system $\{1, 2, 3, \ldots\}$ (cardinals without zero) with the system $\{I, II, III, IV, \ldots\}$ (the same numbers written roman style). They differ greatly in the way in which operations on them are carried out in practice, and yet the operations are in a very real sense the same operations. It is with the similarities emphasized by the isomorphism that algebra is concerned.

## EXERCISES

1. List five members $(u, v)$ of $I(5, 3)$.

2. List five integers $I(u, v)$ which contain $(5, 3)$.

3. Prove, without using Theorem II, that
   a) every member of $I(6, 2)$ is a member of $I(10, 6)$;
   b) every member of $I(10, 6)$ is a member of $I(6, 2)$.

4. Using only the definitions of "+" and "·", and Theorem I, show that

a) $I(2, 5) \text{ "+" } I(7, 3) = I(7, 3) \text{ "+" } I(2, 5)$;

b) $I(2, 5) \text{ "·" } I(7, 3) = I(7, 3) \text{ "·" } I(2, 5)$;

c) $I(2, 5) \text{ "+" } (I(7, 3) \text{ "+" } I(4, 8))$
$= (I(2, 5) \text{ "+" } I(7, 3)) \text{ "+" } I(4, 8)$;

d) $I(2,5)\ "\cdot"\ (I(7,3)\ "\cdot"\ I(4,8))$
$$= (I(2,5)\ "\cdot"\ I(7,3))\ "\cdot"\ I(4,8)\ ;$$

e) $I(2,5)\ "\cdot"\ (I(7,3)\ "+"\ I(4,8))$
$$= (I(2,5)\ "\cdot"\ I(7,3))\ "+"\ (I(2,5)\ "\cdot"\ I(4,8)).$$

5. Prove that the one-to-one correspondence $0 \leftrightarrow 0$, $1 \leftrightarrow 2$, $2 \leftrightarrow 4$, $3 \leftrightarrow 6$, $\cdots$, between the cardinals and the number system of Exercise 9. of Chapter V is not an isomorphism.

6. Prove that the one-to-one correspondence $0 \leftrightarrow I(0,0)$, $1 \leftrightarrow I(0,1)$, $2 \leftrightarrow I(0,2)$, $\cdots$ is not an isomorphism.

7. If (1) $S$ is a set, (2) $*$ is an operation on $S$, and (3) $a$ and $b$ are members of $S$, then we say that $a * b$ is the *composite* of $a$ and $b$. (Thus, in the case that $*$ is an addition operation, the composite is also called the *sum*.)

We now give a definition of isomorphism as a possible relation between any two operations:

Let $S$ be a set with an operation and let $T$ be a set with an operation. Let there be given a one-to-one correspondence between $S$ and $T$ such that the image of the composite of each pair of members of $S$ is the composite of their images. Then the operations are said to be *isomorphic* and the one-to-one correspondence is called an *isomorphism*.

(Thus the definition of isomorphic number systems given in Chapter VI can be restated as follows: Let $M$ and $N$ be number systems such that their addition operations are isomorphic and their multiplication operations are isomorphic under the same one-to-one correspondence. Then $M$ and $N$ are said to be isomorphic number systems.)

a) Prove that the one-to-one correspondence given in Exercise 5 of Chapter VI is an isomorphism of the addition operations.

b) Do the same for Exercise 6 of Chapter VI.

c) Prove that the operation of multiplication on the set considered in Exercise 12 of Chapter VIII is isomorphic to the operation of multiplication on the set of all non-zero integers.

d) Prove that the operation of addition on the set of all non-zero cardinal numbers is isomorphic to the operation of multiplication on the set $\{2^1, 2^2, 2^3, \ldots\}$.

e) Prove that all of the following operations are isomorphic:

| $\cdot$ | 0 | 1 |
|---|---|---|
| 0 | 0 | 0 |
| 1 | 0 | 1 |

| $\cap$ | $\emptyset$ | $\{p\}$ |
|---|---|---|
| $\emptyset$ | $\emptyset$ | $\emptyset$ |
| $\{p\}$ | $\emptyset$ | $\{p\}$ |

| $\cup$ | $\{p\}$ | $\emptyset$ |
|---|---|---|
| $\{p\}$ | $\{p\}$ | $\{p\}$ |
| $\emptyset$ | $\{p\}$ | $\emptyset$ |

(The latter two operations were mentioned in Exercise 7 c) of Chapter V.)   The symbol $\emptyset$ denotes the empty set.

# CHAPTER VII

## PROPERTIES OF THE INTEGERS

### Introduction

IN THIS CHAPTER we derive the chief properties of the number system just constructed, the system of integers. The enterprise is much easier than the corresponding one for the cardinals. This is due to the fact that the whole body of information about generalized laws, expressions, exponents, and polynomials can be carried over with almost no modification into the new number system. The principal novelty comes from the introduction of the notion of the *negative of a number*.

### The Generalized Laws and Expressions

The laws GA, GM, and GD were stated in terms of variables each of whose domains was the set of cardinal numbers. It is obviously possible to formulate similar statements for variables each of whose domains is the set of integers. Moreover, the task of proving such statements true can be carried out very economically. Examination of the proofs of GA, GM, and GD as given for cardinals, shows that they invoke only the fact that the cardinals obey the five fundamental laws. The proofs are therefore valid for any system in which the five fundamental laws are true. This single observation allows us to carry over to integers not only GA, GM, and GD, but also the whole algebra of expressions built up around them.

## The Zero-Element and One-Element of the Integers

We now show that there is an integer that is a zero-element, namely $I(0,0)$.

THEOREM I: *For every integer $I(a,b)$,*

$$I(0,0) + I(a,b) = I(a,b)$$
$$I(0,0) \cdot I(a,b) = I(0,0).$$

*Proof:*  $I(0,0) + I(a,b) = I(0+a, 0+b)$    (definition of addition)

$$= I(a,b) \quad \text{(addition table for cardinals)}$$

$$I(0,0) \cdot I(a,b) = I(0 \cdot b + 0 \cdot a, 0 \cdot a + 0 \cdot b)$$
(definition of multiplication)

$$= I(0,0) \quad \text{(multiplication table for cardinals)}.$$

This proves that $I(0,0)$ has the properties of a zero-element. It is useful to notice that $I(0,0) = I(a,a)$ for any cardinal $a$. This follows from Theorem II of Chapter VI. We shall sometimes designate the zero-element of the integers by the symbol 0, the same symbol that is used to designate the zero-element of the cardinals. While this seems to be inviting confusion, we shall only use this symbol when the context makes clear precisely what we mean.

We now show that there is an integer that is a one-element, namely $I(1,0)$.

THEOREM II: *For every integer $I(a,b)$,*

$$I(a,b) \cdot I(1,0) = I(a,b).$$

*Proof:*  $I(a,b) \cdot I(1,0) = I(a \cdot 1 + b \cdot 0, b \cdot 1 + a \cdot 0)$
(definition of multiplication)

$$= I(a,b) \quad \text{(arithmetic with cardinals)}.$$

This proves that $I(1,0)$ has the property of a one-element. It is useful to note that for any cardinal $a$ we have

$$I(1,0) = I(a+1, a).$$

This follows from Theorem II of Chapter VI. We shall sometimes designate the one-element of the integers by the symbol 1, the same symbol that is used to designate the one-element of the cardinals. Again, we only do this when the context makes clear precisely which we mean.

## The Negative of an Integer

The principal difference between the cardinal number system and the system of integers can be stated in terms of negatives.

THEOREM III: *Every integer $I(a, b)$ has a negative, namely $I(b, a)$.*

*Proof:*  $I(a, b) + I(b,.a)\quad = I(a+b, b+a)$   (by definition of addition)

$$= I(0,0) \qquad \text{(Theorem II, Chapter VI)}.$$

We shall use the symbol $-A$ (read: minus $A$) to designate the negative of $A$. Using this notation, we list and prove some useful facts. They apply not only to the system of integers but to all number systems which have negatives and in which the cancellation laws hold.

(1) $$A + (-A) = 0.$$

This follows from the definition of the negative.

(2) $$-(-A) = A.$$

If $A + B = 0$ then $-B = A$ and $-A = B$. If we substitute for $B$ its equal $-A$ in $-B = A,$ we obtain the result stated.

(3) $$(-1)A = -A.$$

*Proof:* From $1 + (-1) = 0$ and $0 \cdot A = 0$ we have

$$(1 + (-1))A = 0,$$

whence

$$1 \cdot A + (-1) \cdot A = 0.$$

Since

$$1 \cdot A = A,$$

this implies that

$$(-1) \cdot A$$

is the negative of $A$, or

$$(-1) \cdot A = -A.$$

(4) $\quad (-A)B = A(-B) = -(AB)$

*Proof:*
$$\begin{aligned}
(-A)B &= ((-1)A)B && \text{(by (3))}\\
&= (-1)(AB) && \text{(by AM)}\\
&= -(AB) && \text{(by (3))}
\end{aligned}$$

That $A(-B) = -(AB)$ follows from $A(-B) = (-B)(A)$

(5) $$(-A)(-B) = AB.$$

*Proof:*
$$\begin{aligned}
(-A)(-B) &= A(-(-B)) && \text{(by (4))}\\
&= AB && \text{(by (2))}
\end{aligned}$$

(6) $$-(A + B) = (-A) + (-B).$$

*Proof:*
$$\begin{aligned}
-(A + B) &= (-1)(A + B) && \text{(by (3))}\\
&= (-1)A + (-1)B && \text{(by D)}\\
&= (-A) + (-B) && \text{(by (3))}.
\end{aligned}$$

(7) *For each pair of integers $A$ and $B$ there is an integer $X$ such that*

$$X + A = B.$$

*Proof:* $B + (-A)$ has the desired property, since

$$(B + (-A)) + A = B + ((-A) + A) \qquad \text{(AA)}$$
$$= B + 0 \qquad\qquad\qquad ((1))$$
$$= B \qquad\qquad\quad \text{(property of 0)}.$$

## Standard Notation for the Integers

The system of designating integers by symbols such as $I(a, b)$ is not the standard one. In actual practice, the integer that we have called $I(a, 0)$ is written simply as $a$. Thus, not only will we continue to write $I(0,0)$ as 0 and $I(1,0)$ as 1 but we will also write $I(2,0)$ as 2, $I(3,0)$ as 3, etc. In addition, as is customary, we will write $I(0, a)$ as $-a$. This is based on the facts that $I(0,a) = -I(a,0)$ and that $I(a,0) = a$. Thus, we will write $I(0,1)$ as $-1$, $I(0,2)$ as $-2$, etc.

The integers $1, 2, 3, \ldots$ are called the *positive* integers, and the integers $-1, -2, -3, \ldots$ are called the *negative* integers. Note that every integer, whether positive or negative, both has a negative and is itself the negative of some integer. For instance, 5 has $-5$ as its negative and is itself the negative of the integer $-5$. These observations naturally evoke certain questions. How do we designate the remaining integers? How can we tell whether a given symbol like 8 is intended to mean a cardinal or an integer? Why did we introduce the complicated $I(a, b)$ system in the first place if it was to be discarded? The first question is answered by the following theorem:

THEOREM: *For every integer $I(a, b)$ there is an equal integer $I(c, d)$ such that at least one of the cardinals $c$ and $d$ is 0.*

*Examples:* $\qquad I(8, 3) = I(5, 0).$
$\qquad\qquad I(14, 31) = I(0, 17).$

Our answer to the first question, then, is that there are no remaining integers to account for.

*Proof*: If, in $I(a, b)$, either $a$ or $b$ is 0, there is nothing to prove. If neither $a$ nor $b$ is 0, then we can find cardinals $u$ and $v$ such that $a = u + 1$, $b = v + 1$. Then

$$I(a, b) = I(u, v)$$

(Theorem II,<br>Chapter VI).

If either $u$ or $v$ is 0, we have our result. If not, we can find cardinals $p$ and $q$, for which $u = p + 1$, $v = q + 1$. Then, we again have

$$I(u, v) = I(p, q),$$

from which it follows that

$$I(a, b) = I(p, q).$$

This reduction process can be continued until a zero entry is reached. But, since the process replaces the entries by smaller entries, it must eventually stop. Since it can only be stopped by the encountering of a zero entry, it follows that a zero entry must be encountered.

Our answer to the second question is, that it doesn't matter. Consider, for instance, the product $8 \cdot 7$. If the symbols designate cardinals, we can write

$$8 \cdot 7 = 56,$$

where the symbol 56 means the cardinal. The same equation holds if the symbols mean integers. Of course if, in the product $8 \cdot 7$, one symbol means a cardinal and the other an integer, then the expression itself is meaningless, because such multiplications are not defined. What is important here is that $8 \cdot 7 = 56$ is a statement in each of the number systems and is valid in each. This holds true more generally. Given any expression $E$ built up with symbols that represent cardinals, we can construct a corresponding expression involving special integers by the simple device of interpreting each of the number symbols that appears in $E$ as the corresponding special integer. Moreover, the isomorphism between the system of cardinals and the system of special integers insures that if $E$ and $F$ are expressions over the cardinals and if $E = F$ is valid in the cardinal system, then $E = F$ also provides a valid statement about the special integers when the symbols in $E$ and $F$ are reinterpreted as representing the corresponding special integers. This is the content of the statement that there is an isomorphism between the cardinals and the special integers. Our third question has several relevant answers, some of which are obvious. One of the less obvious ones concerns the scope of our operations, which not only constructs the integers from the cardinals but also shows how a new number system can be constructed from any given number system. We could, for instance, consider sets of ordered pairs of integers, treat them as we treated the cardinals, and thus obtain a new number system. It so happens that this would produce a number system indistinguishable from the integers and would therefore be pointless. However, there are number systems other than the cardinals for which this treatment does furnish a significant extension.

## The Cancellation Laws for Integers

The cancellation law for addition of integers states that if for three integers $A$, $B$, $C$ we have

$$A + B = A + C$$

we also have

$$B = C.$$

To prove this, add $-A$ to both members of the first equation

$$(-A) + (A + B) = (-A) + (A + C)$$

and rewrite, using AA

$$((-A) + A) + B = ((-A) + A) + C.$$

It follows that

$$0 + B = 0 + C$$

or

$$B = C.$$

A useful fact about multiplication of integers is that if a product of two integers is zero, one of the factors must be zero. We now prove this.

If, for any integers $A$ and $B$,

$$AB = 0$$

then $\qquad (-A)(-B) = 0 \qquad$ (by (5))

and $\qquad (-A)B = A(-B) = -(AB) \qquad$ (by (4))

$$= -0 = 0.$$

Thus, each of $(-A)(-B), (-A)B, A(-B)$ is zero if $AB$ is zero. Now some one of these four products is composed of positive integers. To evaluate it we use the multiplication table for cardinals, which shows that one of the factors must be zero. Then at least one of the integers $A, -A, B, -B$ is zero, so that at least one of $A, B$ is zero.

The cancellation law for multiplication of integers states that if $A, B, C$ are three integers such that

$$AB = AC$$

then, if $A$ is not zero,

$$B = C.$$

To prove this, observe that if

$$AB = AC$$

then

$$AB + (-(AC)) = AC + (-AC) = 0,$$

from which

$$AB + (A(-C)) = 0, \qquad \text{(from (4))}$$

so that

$$A(B + (-C)) = 0 \qquad \text{(by D)}.$$

Therefore at least one of the integers $A, (B + (-C))$ is zero. We know that $A$ is not. We conclude that

$$0 = B + (-C)$$
$$0 + C = (B + (-C)) + C$$
$$C = B + ((-C) + C)$$
$$= B + 0$$
$$= B.$$

## Expressions Over the Integers

The presence of negatives in the system of integers makes it possible to extend the definition of expression. In dealing with the integers we shall mean by an expression any one of the following:

1) an integer in standard form,
2) a variable whose domain is the set of integers,
3) $(A) + (B)$, if $A$ and $B$ are expressions,
4) $(A)(B)$, if $A$ and $B$ are expressions,
5) $-(A)$, if $A$ is an expression.

*Note*: In 3), 4), 5) no parentheses are to be used around an integer in standard form or around a variable whose domain is the set of all integers.

Two expressions are to be considered equal if they are equal for all values of their variables. We seek a standard form for expressions over the integers. We claim that every such expression is equal to a polynomial. To prove this, we observe that for expressions built up using only 1)-4) this has already been proved. Moreover, the negative of a polynomial is also a polynomial, since

$$-(A + Bx + \ldots + Gx^n)$$
$$= (-1)(A + Bx + \ldots + Gx^n)$$
$$= -A + (-B)x + \ldots + (-G)x^n.$$

Thus any expression whatsoever over the integers equals a polynomial.

## Subtraction

Arithmetic in elementary school is concerned with four operations, addition, subtraction, multiplication, and division. Up to now we have treated only addition and multiplication. Our work with negatives makes it possible to cover the whole subject of subtraction in a few words. The notion of division will be covered similarly in a later chapter.

Definition: Let $N$ be a number system each of whose members has a negative. The *subtraction* operation for $N$ is the operation that assigns to each ordered pair $(A, B)$ of members of $N$ the member $A + (-B)$.

The symbol for this operation is "$-$". Therefore we have

$$A - B = A + (-B).$$

Note that subtraction is neither a commutative nor an associative operation. For instance, in the system of integers, $7 - 5 = 2$, whereas $5 - 7 = -2$, and $7 - (5 - 4) = 6$, whereas $(7 - 5) - 4 = -2$.

This definition does not apply to the system of cardinals, since most of its members do not have negatives. Indeed, while subtraction can be defined for certain pairs of cardinals, it is not an operation on the number system, for it does not assign a cardinal to each ordered pair of cardinals.

The only significance this definition has for our future work is that sometimes, when we add to one number the negative of another, we shall say that we are *subtracting* the second from the first and, what is perhaps more important, we shall omit a pair of parentheses.

## EXERCISES

1. Find the standard form for:

   $2 + 7,\ 2 - 7,\ 7 - 2,\ -2 + 7,\ -7 + 2,\ -7 - 2,\ 2 \cdot 7,$
   $2(-7),\ (-2)(7),\ (-2)(-7),\ -(2 \cdot 7),\ -(2(-7)).$

2. Find the standard form for:

   $2(7 - 3),\ 2(-7 + 3),\ 2(-7 - 3),\ -2(7 + 3),$
   $-2(-7 + 3),\ -2(7 - 3),\ -2(-7 - 3).$

3. Find the standard form for

   $a(b + c),\ a(b - c),\ a(-b + c),\ a(-b - c),\ -a(b + c),$
   $-a(b - c),\ -a(-b + c),\ -a(-b - c).$

4.  Find the standard form for

$$-(x^2 + 2x + 3), \quad -(x^2 - 2x + 3), \quad -(-x^2 - 2x + 3),$$
$$-3(x^2 - 2x - 1), \quad (x + 2)(x^2 - 3x + 4),$$
$$-(x - 2)(x + 3), \quad x(x - 1) - (x + 1)(x + 2).$$

5.  Find a representation of the form $p - q$ for $(a - b) + (c - d)$ and for $(a - b)(c - d)$. Compare the results with $I(a, b)$ "+" $I(c, d)$ and $I(a, b)$ "·" $I(c, d)$.

6.  Show that the sum of a positive integer and a negative integer may be a positive integer, a negative integer, or zero. Give examples to illustrate each case.

7.  Prove that the sum of two negative integers is a negative integer.

8.  Prove that the product of a negative integer and a positive integer is a negative integer.

9.  Prove:
    that the negative of a negative integer is a positive integer;
    that the negative of a positive integer is a negative integer;
    that the negative of zero is zero.

10.  Show that if $A$ is an integer which is such that $A = -A$, then $A = 0$.

11.  Is it always true that the negative of an integer is a negative integer?

12.  Let $C$ denote the set of all cardinal numbers. We now define two operations on the set $C \times C$. Let $(a, b)$ and $(c, d)$ be any members of $C \times C$. We define as their "product" and designate by the symbol $(a, b) \odot (c, d)$ the member $(ac, bd)$. We define as their "sum" and designate by the symbol $(a, b) \oplus (c, d)$ the member $(a + c, b + d)$.

a)  Show that the set $C \times C$ with these operations is a number system.

b)  Does this number system have a zero-element? a one-element?

c)  Does this number system have the property proved for the cardinal number system in Theorem XI of Chapter II?

d) Show by a suitable example that this number system does not have a cancellation law for multiplication.

e) Show that under these operations the subset $\{(0,0), (1,0), (2,0), (3,0), \ldots\}$ of $C \times C$ is a number system that is isomorphic to the cardinal number system. Notice that $(1,0)$ is a one-element with respect to this subset but not with respect to $C \times C$.

# CHAPTER VIII

## THE RATIONAL NUMBER SYSTEM

### Introduction

THE NUMBER SYSTEM to be constructed in this chapter is called the *rational number system*. The methods that we employ in its construction are similar to those used in the construction of the integers. We first assemble a set, define two operations on the set, and then show that the given set and the given operations combine to form a number system. We then examine this system of rational numbers in detail. It is shown to be an extension of the system of integers having all the important properties of that number system and a new feature of its own.

### The Set of Rational Numbers

Our first step in constructing the rational number system is to assemble its members. To this end we consider the set of all ordered pairs $(a, b)$, where $a$ is an integer and $b$ is an integer other than zero. We designate this set by the symbol $T$. The rational numbers are defined as certain subsets of this set $T$.

DEFINITION: For each member $(a, b)$ of $T$, $R(a, b)$ is the set of all members $(u, v)$ of $T$ for which $av = bu$ holds. We call each such subset $R(a, b)$ of $T$ a *rational number*.

*Example*: $(15, -3)$ is a member of the subset $R(-5, 1)$, because $(-5)(-3) = (1)(15)$. $(14, -2)$ is not a member of $R(-5, 1)$, because $(-5)(-2) \neq (1)(14)$. We need a few facts

111

about these subsets $R(a, b)$. Note first that every member $(p, q)$ of $T$ is a member of such a subset, namely $R(p, q)$, We verify this by applying to $(p, q)$ the test for membership in $R(p, q)$, which is that $pq = qp$. This follows from the fact that $p$ and $q$ are integers and obey the law CM. We also prove that there is only one subset $R(a, b)$ that contains a given member $(p, q)$ of $T$.

THEOREM I: *If $R(p, q)$ and $R(s, t)$ have a common member $(u, v)$, then $R(p, q) = R(s, t)$.*

*Note:* We have already shown that $R(-5, 1)$ and $R(15, -3)$ have $(15, -3)$ as a common member. Our theorem implies that $R(-5, 1)$ and $R(15, -3)$ are the same. Of course, the symbols "$R(-5, 1)$" and "$R(15, -3)$" are different. The point is that they must designate the same subset of $T$; that is, any member of one must also be a member of the other.

*Proof:* Let $(l, m)$ be a member of $R(p, q)$. We show that it must also be a member of $R(s, t)$. We have

| | |
|---|---|
| $pm = ql$ | ($(l, m)$ is in $R(p, q)$) |
| $pv = qu$ | ($(u, v)$ is in $R(p, q)$) |
| $sv = tu$ | ($(u, v)$ is in $R(s, t)$) |
| $pmqusv = qlpvtu$ | (by multiplication) |
| $(pquv)sm = (pquv)tl$ | (GM). |

Now if $pquv$ is not zero, we apply the cancellation law for multiplication of integers to obtain

$$sm = tl,$$

which proves that $(l, m)$ is in $R(s, t)$. If $pquv$ is zero, we cannot invoke this law. However, we do know that neither $q$ nor $v$ can be zero, since we are dealing with members of $T$, so if $pquv$ is zero, $pu$ must be zero. Returning to $pv = qu$, we see that both $p$ and $u$ must then be zero. From $pm = ql$ we then have that $l$ is zero, and from $sv = tu$ we have that $s$ is zero. It follow that $sm$ and $tl$ are both zero, whence $sm = tl$. And here, too, we have established that $(l, m)$ is in $R(s, t)$. Thus it follows that every member of $R(p, q)$ is a member of $R(s, t)$. Similarly, it can be shown that every member of $R(s, t)$ is a member of $R(p, q)$. This shows that $R(p, q)$ and $R(s, t)$ are the same.

We have shown that if two rationals have a single member in common, they must be the same. The other possible relation between two rationals is that they have no members in common.

It is useful to have a test to decide whether two given rationals are the same. The following theorem provides such a test.

Theorem II: *If $R(p, q) = R(s, t)$, then $pt = qs$. Conversely, if $pt = qs$, then $R(p, q) = R(s, t)$.*

*Proof:* If $R(p, q) = R(s, t)$, then every member of $R(s, t)$ is a member of $R(p, q)$. But we know that $(s, t)$ is such a member. It must therefore satisfy the test for membership in $R(p, q)$, which is $pt = qs$. This proves the first part of the theorem. To prove the second part, we deduce from $pt = qs$ that $(s, t)$ belongs to $R(p, q)$. We know that $(s, t)$ belongs to $R(s, t)$. We conclude that $R(s, t)$ and $R(p, q)$ have a common member and are then equal, by Theorem I.

## Addition and Multiplication of Rationals

We now define two operations on the set of rationals which we provisionally designate by the symbols " + " and " · ".

Definitions:

$$R(x, y) \text{ "+" } R(u, v) = R(xv + yu, yv)$$
$$R(x, y) \text{ "·" } R(u, v) = R(xu, yv),$$

where $x$ and $u$ are variables whose domain is the set of integers and where $y$ and $v$ are variables whose domain is the set of all integers except zero.

*Note:* Each of these two definitions is an abbreviation for an endless list of individual definitions, such as

$$R(3, -2) \text{ "+" } R(-6, 8) = R(36, -16)$$

and

$$R(3, -2) \text{ "·" } R(-6, 8) = R(-18, -16).$$

The following theorem asserts that the rationals, under " + " as addition and " · " as multiplication, constitute a number system.

THEOREM III: *Each of the operations "+" and "·" is commutative and associative, and "·" is distributive with respect to "+".*

*Proof of* (CA): To show that "+" is commutative, we must prove that

$$R(a, b) \text{ "+" } R(c, d) = R(c, d) \text{ "+" } R(a, b).$$

We have

$$R(a, b) \text{ "+" } R(c, d) = R(ad + bc, bd)$$

and

$$R(c, d) \text{ "+" } R(a, b) = R(cb + da, db).$$

These two are equal, by CM and CA for integers.

*Proof of* (CM): To show that "·" is commutative, we must prove that

$$R(a, b) \text{ "·" } R(c, d) = R(c, d) \text{ "·" } R(a, b).$$

We have

$$R(a, b) \text{ "·" } R(c, d) = R(ac, bd)$$

and

$$R(c, d) \text{ "·" } R(a, b) = R(ca, db).$$

These two are equal, by CM for integers.

*Proof of* (AA): To show that "+" is associative, we must prove that

$$R(a, b) \text{ "+" } (R(c, d) \text{ "+" } R(e, f))$$
$$= (R(a, b) \text{ "+" } R(c, d)) \text{ "+" } R(e, f)$$

We have

$$R(a, b) \text{ "+" } (R(c, d) \text{ "+" } R(e, f))$$
$$= R(a, b) \text{ "+" } R(cf + de, df)$$
$$= R(adf + b(cf + de), bdf)$$
$$= R(adf + bcf + bde, bdf).$$

Similarly,

$$(R(a, b) \text{ "+" } R(c, d)) \text{ "+" } R(e, f)$$
$$= R(ad + bc, bd) \text{ "+" } R(e, f)$$
$$= R((ad + bc)f + bde, bdf)$$
$$= R(adf + bcf + bde, bdf).$$

Thus the two expressions are equal.

*Proof of* (AM): To show that "$\cdot$" is associative, we must prove that

$$R(a,b) \text{``}\cdot\text{''} (R(c,d) \text{``}\cdot\text{''} R(e,f))$$
$$= (R(a,b) \text{``}\cdot\text{''} R(c,d)) \text{``}\cdot\text{''} R(e,f).$$

We have

$$R(a,b) \text{``}\cdot\text{''} (R(c,d) \text{``}\cdot\text{''} R(e,f))$$
$$= R(a,b) \text{``}\cdot\text{''} R(ce,df)$$
$$= R(ace,bdf).$$

Similarly,

$$(R(a,b) \text{``}\cdot\text{''} R(c,d)) \text{``}\cdot\text{''} R(e,f)$$
$$= R(ac,bd) \text{``}\cdot\text{''} R(e,f)$$
$$= R(ace,bdf).$$

This proves our result.

*Proof of* (D): To show that "$\cdot$" is distributive relative to "$+$", we must prove that

$$R(a,b) \text{``}\cdot\text{''} (R(c,d) \text{``}+\text{''} R(e,f))$$
$$= (R(a,b) \text{``}\cdot\text{''} R(c,d)) \text{``}+\text{''} (R(a,b) \text{``}\cdot\text{''} R(e,f))$$

We have

$$R(a,b) \text{``}\cdot\text{''} (R(c,d) \text{``}+\text{''} R(e,f))$$
$$= R(a,b) \text{``}\cdot\text{''} R(cf + de, df)$$
$$= R(acf + ade, bdf).$$

We also have

$$(R(a,b) \text{``}\cdot\text{''} R(c,d)) \text{``}+\text{''} (R(a,b) \text{``}\cdot\text{''} R(e,f))$$
$$= R(ac,bd) \text{``}+\text{''} R(ae,bf)$$
$$= R(acbf + bdae, b^2 df)$$
$$= R(b(acf + ade), b(bdf)).$$

It follows from Theorem II that these two expressions are equal.

This theorem shows that the rationals, under the operations "$+$" and "$\cdot$", constitute a number system whose addition operation is the operation "$+$" and whose multiplication operation is the operation "$\cdot$". We shall from now on write "$+$" and "$\cdot$" without quotation marks. We shall also use the laws GA, GM, GD in the rational number system. Our justification is that any number system that obeys the five fundamental laws obeys these generalized laws.

## One-Element, Zero-Element, and Negatives

We show that the rational number system has a one-element, namely, $R(1, 1)$. To verify this, we must show that for all rationals $R(a, b)$,

$$R(1, 1) \cdot R(a, b) = R(a, b).$$

We have

$$R(1, 1) \cdot R(a, b) = R(1 \cdot a, 1 \cdot b) \quad \text{(definition of multiplication)}$$

$$= R(a, b) \quad \text{(properties of the integer 1)}.$$

It is useful to observe that the one-element for the rationals is also designated by $R(2, 2)$, $R(-7, -7)$ and by every symbol of the form $R(a, a)$. To prove this—that $R(a, a) = R(1, 1)$—we need only apply Theorem II.

We show that the rational number system has a zero-element, namely $R(0, 1)$. We are to show that for each rational $R(a, b)$

$$R(0, 1) + R(a, b) = R(a, b)$$
$$R(0, 1) \cdot R(a, b) = R(0, 1).$$

We first observe that a rational of the form $R(0, q)$, where $q$ is any non-zero integer, equals $R(0, 1)$. This follows from Theorem II. A proof of the first property is

$$R(0, 1) + R(a, b) = R(0 \cdot b + 1 \cdot a, 1 \cdot b) \quad \text{(definition of addition)}$$

$$= R(a, b).$$

A proof of the second property is

$$R(0, 1) \cdot R(a, b) = R(0 \cdot a, 1 \cdot b) \quad \text{(definition of multiplication)}$$

$$= R(0, b)$$
$$= R(0, 1) \quad \text{(by above remark)}.$$

We now show that every rational has a negative, and more specifically, that the negative of $R(a, b)$ is $R(-a, b)$. We have

$$R(a, b) + R(-a, b) = R(ab + b(-a), b^2) \quad \text{(definition}$$
$$\text{of addition)}$$
$$= R(ab - ba, b^2)$$
$$= R(0, b^2)$$
$$= R(0, 1).$$

### Relation Between Rationals and Integers

We have completed our construction of the rational number
system and have derived some of its properties. One of its most
important properties is that it is an extension of the integers:

THEOREM IV: *The correspondence that assigns to each integer $p$
the rational $R(p, 1)$ is an isomorphism.*

*Proof*: To prove this theorem, we must establish two facts. First,
we must show that, if $p$ and $q$ are any integers, the rational that
corresponds to $p + q$ is the sum of the rationals that correspond
to $p$ and to $q$. The second is that, with $p$ and $q$ as above, the rational
that corresponds to $pq$ is the product of the rationals that correspond to $p$ and to $q$. In symbols, this means we must show that

and
$$R(p + q, 1) = R(p, 1) + R(q, 1)$$
$$R(pq, 1) = R(p, 1) \cdot R(q, 1).$$

That the first of these is valid follows immediately from the definition of addition of rationals, and that the second is valid follows
immediately from the definition of multiplication of rationals. Thus
the theorem is proved.

This brings us to a matter of notation. The $R(x, y)$ system for
designating rationals is not the customary one. It is traditional to
designate what we have been calling $R(14, 1)$ by the symbol 14,
and in general, what we have been calling $R(a, 1)$, by the symbol $a$.
The same symbol thus is used for an integer and its corresponding

rational, so that it is not possible to tell whether the expression $2 + 2$ is the cardinal 4, the integer 4, or the rational 4. Fortunately, because of the isomorphisms, the sum $2 + 2$ is 4 in each of these number systems, so that no difficulty arises provided one operates in a single number system.

We shall not yet abandon the $R(x, y)$ notation for the more usual scheme for designating rationals. We have used it for theoretical purposes, and one major bit of theory remains to be derived.

## Reciprocals

The whole system of integers is isomorphic to some of the rationals. Let us see what the presence of "extra" rationals implies. Our conclusions can be stated in terms of reciprocals.

THEOREM V: *Every rational $R(a, b)$ except the zero element has a reciprocal. The reciprocal of such an $R(a, b)$ is $R(b, a)$.*

*Proof*: Since $R(a, b)$ is not the zero element, $a$ is not zero. It follows that $R(b, a)$ is a legitimate symbol for a rational, and

$$R(a, b)R(b, a) = R(ab, ba).$$

Since $R(ab, ba) = R(1, 1)$ by Theorem II, we have shown that $R(a, b)$ has a reciprocal, namely $R(b, a)$.

If $R(a, b)$ is the zero-element, then it has no reciprocal. This follows from the fact that then $R(a, b) \cdot R(x, y)$ is the zero-element for every $R(x, y)$ and therefore cannot equal the one-element, as required by the definition of reciprocal, for any $R(x, y)$.

Note that the only cardinal that has a reciprocal is 1, and the only integers that have reciprocals are 1 and $-1$. Thus the rationals are quite different from the cardinals and the integers as regards reciprocals.

## Traditional Notation for Rationals

The notation for rationals in general usage designates what we have called $R(x, y)$ by the symbol $x/y$ or by the symbol $\dfrac{x}{y}$. In this notation, our definitions become

$$\frac{x}{y} + \frac{u}{v} = \frac{xv + yu}{yv}$$

$$\frac{x}{y} \cdot \frac{u}{v} = \frac{xu}{yv}.$$

Since the reciprocal of $R(x, y)$ is $R(y, x)$, the reciprocal of $x/y$ is $y/x$. In particular, the reciprocal of $R(y, 1) = y/1$ is $R(1, y) = 1/y$. Thus, from the equation

$$\frac{x}{1} \cdot \frac{1}{y} = \frac{x}{y},$$

we can conclude that every rational number is the product of a special rational $x/1 = x$ by the reciprocal of a special rational $y/1 = y$. It is usual to call the expression $x/y$ a *fraction* and to call $x$ the numerator and $y$ the denominator. From what has been said so far, a symbol as $x/y$ has meaning if $x$ and $y$ are integers and $y$ is different from zero. We also give a meaning to symbols such as $A/B$, where $A$ and $B$ are rationals, with $B$ not zero. It is to mean the product of $A$ by the reciprocal of $B$. Thus $(3/4)/(5/9 + 1/7)$ means $3/4$ multiplied by the reciprocal of $(5/9 + 1/7)$, that is, by the reciprocal of $44/63$, which is $63/44$. Thus

$$\frac{3/4}{5/9 + 1/7} = \frac{3}{4} \cdot \frac{63}{44} = \frac{189}{176}.$$

This is perhaps a suitable place to state that to divide $A$ by $B$ is to be understood as meaning the result of multiplying $A$ by the reciprocal of $B$. Division requires no separate lore. It is handled by reference to multiplication and the finding of reciprocals.

## Standard Form

A rational number can be designated by many different symbols. For each rational number, we single out one particular symbol and call it the *standard form* of that number.

We first need a few definitions and facts about integers. We say that an integer $p$ is a *factor* of an integer $a$ if there is an integer $q$ for which $a = pq$. We say that an integer $p$ is a *common factor* of the integers $a$ and $b$ if it is a factor of each. We say that an integer $p$ is a *highest common factor* of the integers $a$ and $b$ if $p$ is a common factor of $a$ and $b$ and if every common factor of $a$ and $b$ is a factor of $p$. It can be shown that every pair of integers $a, b$ at least one of which is not zero has exactly two highest common factors, each the negative of the other. We illustrate these remarks with examples.

2 is a factor of the integer 12. $-2$ is a common factor of the pair of integers 12 and 8; 2, $-1$, 1, $-4$, and 4 are other common factors. 4 and $-4$ are the highest common factors of the pair 12 and 8. We can now give a procedure for selecting a standard form for each rational. We use the $R(x, y)$ notation. Let $R(a, b)$ designate a rational. Recall that $a$ and $b$ are integers and that $b$ is not zero. Let $p$ be a highest common factor of $a$ and $b$, so that $a = up$, $b = vp$, where $u$ and $v$ are integers whose highest common factor is 1. We assert that

$$R(a, b) = R(u, v) = R(-u, -v),$$

and this follows from Theorem II and the equations $a = up, b = vp$. We select, of $R(u, v)$ and $R(-u, -v)$, that one whose second entry is positive as the standard form for $R(a, b)$. For instance, the standard form of $R(12, -8)$ is $R(-3, 2)$ because, in the above notation, $a = 12$, $b = -8$, $p = 4$, $u = 3$, $v = -2$; and $R(-u, -v)$, rather than $R(u, v)$, makes the second entry positive.

## Expressions Over the Rationals

The presence of reciprocals in the rational number system makes it possible to extend the definition of expression. When dealing with the rational number system, we shall mean by an expression any one of the following:

(1) a rational number in standard form,
(2) a variable whose domain is the set of rationals,
(3) $(A) + (B)$, where $A$ and $B$ are expressions,
(4) $(A)(B)$, where $A$ and $B$ are expressions,
(5) $-(A)$, where $A$ is an expression,
(6) $A/B$, where $A$ is an expression, and $B$ is an expression that is not equal to zero.

*Note*: It is understood that in (3), (4), and (5) no parentheses are to be used around a rational that is in standard form or around a variable whose domain is the set of rationals.

It is only item (6) that is novel in this list. Let us first consider those expressions formed without invoking item (6). By arguments precisely similar to those used before, we can show that all the rules and all the conclusions about standard forms which we have already deduced apply to these expressions as well. Thus in the study of expressions over the rationals, we need only examine the novelties introduced by (6) above. It commits us to considering combinations of symbols such as $1/x$, $(3x - 7)/(x^2 + 2x + 3)$ as well as more complicated ones. It turns out that any expression formed over the rationals has one equal to it of the form $A/B$, where $A$ and $B$ are polynomials in standard form. We are going to show how to find such expressions and how to select standard forms from them.

We must first reexamine the notion of equal expressions. Our original definition was a simple application of our agreement about statement-forms and variables. Thus, $(x + 1)^2 = x^2 + 2x + 1$ was considered as meaningless unless a domain was specified for the variable; and if a domain was specified, all the statements obtained by substituting each member of the domain for $x$ were to be asserted jointly. This is no longer possible with the rationals, because a perfectly legitimate expression may not admit such total substitution. For example, we cannot substitute $-1$ for $x$ in $1/(x + 1)$, because this would create a fraction with a zero denominator, which is a forbidden object. Again, the expressions $x$

and $x^2/x$ are equal for all values of $x$ except zero; but zero cannot be substituted into the second expression. We get around these difficulties by agreeing that two expressions are to be called equal if their values are equal for all substitutions of their variables by those rationals that create a zero denominator in neither of them. From this point of view we can, for instance, write $x = x^2/x$. It cannot be that all possible substitutions yield a zero denominator, because such a denominator would equal zero and hence would be ruled out by (6).

It follows immediately that

$$\frac{A}{B} + \frac{C}{D} = \frac{AD+BC}{BD} , \quad \frac{-A}{B} = \frac{A}{-B} = -\frac{A}{B} , \left(\frac{A}{B}\right)\left(\frac{C}{D}\right) = \frac{(AC)}{(BD)}$$

and that

$$A \cdot (1/A) = 1.$$

We now take up the problem of finding a standard form for an expression $E$. We first show that there is an equal expression of the form $A/B$, where $A$ and $B$ are polynomials. To this end, let us follow out the construction of $E$ step by step. As long as the steps do not invoke item (6) above, what is constructed equals a polynomial. Thus, what results from the first use of (6) equals a fraction whose numerator and denominator are both polynomials. Further steps may require the construction of other such fractions and the construction of their sums, products, negatives, and reciprocals. The rules listed in the first paragraph of this section guarantee that any expressions so constructed also equal a fraction whose numerator and denominator are polynomials. This completes our argument to show that every expression equals a fraction whose numerator and denominator are polynomials.

To select the standard forms for such expressions, we follow the procedure, used for the rational numbers themselves, of requiring that the fractions be in lowest terms: we will say that $A/B$, where $A$ and $B$ are polynomials, is in *lowest terms* if there is no polynomial $C$ of positive degree for which $A = PC$, $B = QC$, where $P$ and $Q$ are polynomials. If there are such polynomials $C$, $P$, and $Q$, then clearly $A/B = P/Q$. Thus, $x^2/x$ is not in lowest terms but it is equal to an expression $x = x/1$ which is in lowest terms. We claim that every expression $A/B$, where $A$ and $B$ are polynomials, is equal to an expression $P/Q$, where $P$ and $Q$ are polynomials and $P/Q$ is in lowest terms. We accept any such expression as a standard form. We recognize that an expression may have several such standard forms; for instance $x = 2x/2 = (3/5)x/(3/5)$. The notion of standard form can be refined further so that each expression has only one standard form. In the polynomial case we actually achieved this uniqueness of standard form, having in mind our later work on equations. In the present case, nothing comparable is at stake, so we do not take up this further refinement.

## EXERCISES

1. List five members $(u, v)$ of $R(5,3)$.

2. List five rationals $R(u, v)$ that contain $(5, 3)$.

3. Prove, without using Theorem II, that
   a) every member of $R(4, 10)$ is a member of $R(6, 15)$;
   b) every member of $R(6, 15)$ is a member of $R(4, 10)$.

4. Using only the definitions of "+" and "·", and Theorem I, show that

   a) $R(2, 5)\,"+"\,R(7, 3) = R(7, 3)\,"+"\,R(2, 5)$;

   b) $R(2, 5)\,"·"\,R(7, 3) = R(7, 3)\,"·"\,R(2, 5)$;

   c) $R(2, 5)\,"+"\,(R(7, 3)\,"+"\,R(4, 8))$
   $\qquad = (R(2, 5)\,"+"\,R(7, 3))\,"+"\,R(4, 8)$;

   d) $R(2, 5)\,"·"\,(R(7, 3)\,"·"\,R(4, 8))$
   $\qquad = (R(2, 5)\,"·"\,R(7, 3))\,"·"\,R(4, 8)$;

   e) $R(2, 5)\,"·"\,(R(7, 3)\,"+"\,R(4, 8))$
   $\qquad = (R(2, 5)\,"·"\,R(7, 3))\,"+"\,(R(2, 5)\,"·"\,R(4, 8))$.

5. Find the standard form for

   $2/5 + 7/3,\ 2/5 - 7/3,\ -2/5 + 7/3,\ -2/5 - 7/3$.

6. Find the standard form for

$$-(2/5 + 7/3),\ -(2/5 - 7/3),$$
$$-(-2/5 + 7/3),\ -(-2/5 - 7/3).$$

7. Find the standard form for

$$\frac{1}{2/5},\ \frac{3}{2/5},\ \frac{3/2}{5},\ \frac{1}{2/5 - 7/3},\ \frac{4/5 + 5/6}{7/3 - 2/5}.$$

8. Find the standard form for

$$1/2 + 1/b,\ 2/b + b/a,\ b/a + c/a,\ b/a + c/a^2.$$

9. Find the standard form for

$$\left(\frac{x}{x+1}\right)^2,\ \frac{x}{x+1} + \frac{x^2 + 2x}{4x - 5},\ \frac{2x - 3}{x + 3} + 2.$$

10. Show that $a/c + b/c = (a + b)/c$ if $c$ is not zero, where

a) $a$, $b$, and $c$ are integers;
b) $a$, $b$, and $c$ are rationals.

11. Show that $ab/ac = b/c$ if $a$ and $c$ are not zero, where

a) $a$, $b$, and $c$ are integers;
b) $a$, $b$, and $c$ are rationals.

12. Show that the rationals of the form $R(1, a)$, where $a$ is a non-zero integer, do not form a number system relative to the operations of addition and multiplication defined originally for the set of all rational numbers.

13. Find the standard form for

$$\frac{2}{3} + \frac{4}{5},\qquad \frac{2+4}{3+5},\qquad \frac{2}{3} \cdot \frac{4}{5},\qquad \frac{2 \cdot 4}{3 \cdot 5},\qquad \frac{\frac{2}{3}}{4},\qquad \frac{2}{\frac{3}{4}}.$$

# CHAPTER IX

## EQUATIONS

### Introduction

IN THIS CHAPTER we conclude our work on the problem of deciding if two given expressions are equal; we study the values of the variables for which given expressions have equal values. This broadens our original investigation, which was concerned with pairs of expressions having equal values for all values of their variables. By tackling the more general problem, we not only solve the original one but also make contact with a part of algebra that has numerous practical applications. It often happens that in an investigation remote from mathematics questions arise that can be rephrased as questions about the values of variables for which certain expressions have equal values. Results of the sort contained in this chapter can then furnish useful information about the answers. Unfortunately, most of these questions require a special technical training in some other discipline, and so we cannot consider them here. We can, however, give the reader some indication of these applications of mathematics by solving a few more-or-less simple puzzles along these lines.

### Definitions

We have often considered statement-forms $E = F$, where $E$ and $F$ are expressions. It was understood that the variables appearing in the expressions were to be assigned some number system as domains, and $E = F$ was then to mean that the value of $E$ was the same as the value of $F$ for each value of the variables. We now introduce a new interpretation for $E = F$. We shall sometimes consider $E = F$ to be a statement in which the variables have as their domain *that subset of a specified number system for which the associated statements are true*. With this interpretation, we shall say that $E = F$ is an *equation*.

Consider, for example, $(x + 1)^2 = x^2 + 1$; this is, in our old interpretation, a false statement if $x$ is a variable whose domain is the set of rationals. Our new interpretation of $(x + 1)^2 = x^2 + 1$ is one that regards $x$ as a variable whose domain is that subset of the rationals for which the associated statements are true. Thus, in this interpretation, $(x + 1)^2 = x^2 + 1$ is automatically true, even though we may not know the domain of the variable $x$. The members of such a domain are called *solutions* of the equation. More precisely, letting $E$ and $F$ be expressions constructed with members of a number system $N$ and with variables $x, y, \ldots, w$, a solution of the equation $E = F$ is a set of numbers $a, b, \ldots, g$ of $N$, one for each variable, which when substituted for their variables in $E$ and $F$ produce equal values for $E$ and for $F$.

A source of possible ambiguity in dealing with equations lies in the specification of the number system to be used. Sometimes no specification at all is given. Under that circumstance, the expressions that figure in the equations furnish some clue, because only those number systems could reasonably be intended which contain the numbers that appear in these expressions. Unfortunately, this does not settle the matter completely. We can illustrate what is involved here by considering a famous equation, $x^2 + 1 = 0$. For many years it was debated whether or not this equation actually had solutions. It turned out that both those who claimed it did and those who claimed it did not were correct, provided a suitable intepretation was put on the claims. Those who claimed it did not have solutions were correct if they meant that no number system familiar to them contained a solution of the equation. Those who claimed that it did have solutions were correct if they meant that a new number system could be constructed containing a one element and a zero element and also a number $a$ for which $a^2 + 1 = 0$. (The one and zero elements are needed in order to have $a^2 + 1 = 0$ make sense.) It should be pointed out that our new interpretation of $E = F$ contains the earlier one as a special case. For instance, $(x + 1)^2 = x^2 + 2x + 1$ can be regarded either as a statement of the equality of two expressions in the previous sense or as an equation whose solutions are the whole number system. Sometimes the term *identity* is used when the first interpretation is to be employed, and *equation* otherwise. We shall not use this terminology but shall try to make our meaning clear in each case as it arises.

## Classification and Solution of Equations

A fundamental problem in the study of equations is to determine their solutions. One way of doing this is to deduce from the given equation another equation which has the same solutions but is handled more easily. For instance, $(x + 1)^2 = x^2 + 1$ has the same solutions as $2x = 0$, and the number zero is clearly the sole solution of $2x = 0$. Three frequently-used devices in this connection are

(1) to replace $A = B + C$ by $A - C = B$ (accomplished by adding $- C$ to both sides).

(2) to replace $A = BC$ by $A/C = B$ if $C \neq 0$ (accomplished by multiplying both sides by $1/C$).

(3) to infer from $AB = 0$ that either $A = 0$ or $B = 0$.

By the use of these devices, an equation can sometimes be replaced by another, with the same solutions, of the form $P = 0$, where $P$ is a polynomial in standard form. Such equations are classified according to the degree of the polynomial $P$ and the number of variables in $P$. We shall study for the most part equations involving one variable. We shall obtain an estimate of the number of solutions of such equations, and for equations of degree one and two shall find the actual solutions.

## Equations in One Variable

We now prove three theorems that give information about the number of solutions of an equation in one variable and that enable us to settle our problem about the equality of polynomials in standard form. The theorems are stated without reference to any particular number system; the proofs are valid for any number system in which every number has a negative, and in which (3) of the preceding section is valid.

THEOREM I: *Let n be a positive integer, and let A and B be expressions. Then*

$$A^n - B^n = (A - B)(A^{n-1} + A^{n-2}B + \ldots + B^{n-1}),$$

*where the sum in the second parentheses contains as a term each product $A^p B^q$ for which p and q are positive integers whose sum is $n - 1$.*

*Examples*: The theorem states, for example, that

$$A^2 - B^2 = (A - B)(A + B),$$
$$x^3 - y^3 = (x - y)(x^2 + xy + y^2),$$
$$p^4 - 1 = (p - 1)(p^3 + p^2 + p + 1).$$

*Proof*: We have

$$(A - B)(A^{n-1} + A^{n-2}B + \ldots + B^{n-1})$$
$$= A^n + A^{n-1}B + \ldots + AB^{n-1}$$
$$\quad - A^{n-1}B - \ldots - AB^{n-1} - B^n$$
$$= A^n - B^n.$$

THEOREM II (Factor theorem): *Let n be a positive integer, $P(x)$ a polynomial of degree n, r a number for which $P(r) = 0$. Then there is a polynomial $Q(x)$ of degree $n - 1$ which is such that*

$$P(x) = (x - r)Q(x).$$

*Example*: Let $n = 3$, $P(x) = 2x^3 - x^2 + x - 2$ and let $r = 1$. A simple calculation shows that $P(1) = 0$ and that $P(x) = (x - 1)(2x^2 + x + 2)$.

*Proof*: Suppose that

$$P(x) = ax^n + bx^{n-1} + \ldots + px + q,$$

where $a, b, \ldots, p, q$ are numbers and $a$ is not zero. We have

$$P(r) = ar^n + br^{n-1} + \ldots + pr + q,$$

whence

$$P(x) - P(r)$$
$$= a(x^n - r^n) + b(x^{n-1} - r^{n-1}) + \ldots + p(x - r).$$

By Theorem I, each term of this expression has a representation as the product of $(x - r)$ by a certain polynomial. Thus, by law GD, the whole expression has a representation as the product of $(x - r)$ by the sum of these polynomials. In symbols,

$$a(x^n - r^n) = (x - r)(ax^{n-1} + \ldots + ar^{n-1})$$
$$b(x^{n-1} - r^{n-1}) = (x - r)(bx^{n-2} + \ldots + br^{n-2})$$
. . . . . . . . . . . . . . . . . . . . . . . . . . . . . . . . . . . . . . . . . . . .

$$p(x - r) = (x - r)p$$

and

$$P(x) - P(r) = (x - r)(ax^{n-1} + \ldots + p)$$

and finally, since $P(r) = 0$,

$$P(x) = (x - r)(ax^{n-1} + \ldots + p),$$

and this is what was to be shown.

*Note*: The converse of this theorem, namely, that if

$$P(x) = (x - r)Q(x),$$

then $r$ is a solution of $P(x) = 0$ is easy to prove, for

$$P(r) = (r - r)Q(r) = 0.$$

This fact can occasionally be used to solve equations. For instance it is possible, with some ingenuity, to deduce that $x^2 - 3x + 2 = (x - 2)(x - 1)$ from which it follows that $x = 1$ and $x = 2$ are solutions of $x^2 - 3x + 2 = 0$.

THEOREM III: *Let $n$ be a positive integer, and let $P(x)$ be a polynomial of degree $n$. Then there are at most $n$ distinct numbers $r, s, t, \ldots, u$ for which*

$$P(r) = 0, P(s) = 0, P(t) = 0, \ldots, P(u) = 0.$$

In other words: *An equation of degree $n$ has at most $n$ solutions.*

It may well have fewer than $n$ solutions. For instance, the equation $x^9 = 0$, which is of degree nine, has only one solution.

*Proof:* By the factor theorem, we have

$$P(x) = (x-r)Q(x),$$

where $Q(x)$ is a polynomial of degree $n-1$. Since $P(s) = 0$

$$(s-r)Q(s) = 0,$$

and since $s-r$ is not zero ($r$ and $s$ are distinct) we have $Q(s) = 0$. Again, by the factor theorem, we have

$$Q(x) = (x-s)R(x),$$

where $R(x)$ is a polynomial of degree $n-2$. It follows that

$$P(x) = (x-r)(x-s)R(x).$$

Then

$$P(t) = (t-r)(t-s)R(t)$$

and, since $P(t) = 0$ and $(t-r)(t-s)$ is not zero ($r$, $s$, $t$ are distinct), we have $R(t) = 0$. Again, by the factor theorem, we have

$$P(x) = (x-r)(x-s)(x-t)S(x).$$

Continuing in this way, if we have as many as $n$ distinct numbers $r, s, t, \ldots, u$ which make $P(x)$ zero, we have

$$P(x) = (x-r)(x-s)(x-t)\ldots(x-u)V$$

where, since the degree of $V$ is obtained by subtracting one $n$ times in all from $n$, $V$ is a polynomial of degree zero. In other words, $V$ is some number other than zero. We are to show that if $w$ is also a solution of $P(x) = 0$, then $w$ must be one of $r, s, \ldots, u$. We have

$$P(w) = (w-r)(w-s)\ldots(w-u)V,$$

and $P(w)$ can not be zero unless one of its factors is zero, which can only happen if $w$ is some one of $r, s, \ldots, u$.

COROLLARY : *If two polynomials in standard form having rational coefficients are equal, then each term of one must be a term of the other.*

*Proof*: Let $P$ and $Q$ be two equal polynomials in standard form. Their equality means that for every value of $x$ the value of $P - Q$ is zero. Now if $P$ had a term $ax^r$ not in $Q$, the polynomial $P - Q$ would have the term $ax^r$ and therefore, not being the zero polynomial, would have some degree $n$. But by the last theorem, $P - Q$ could then have the value zero for no more than $n$ numbers and certainly not for all the infinitely many rationals. Thus, either $P$ and $Q$ are not equal or they have the same terms.

*Note*: While this corollary was proved for the case of the rational number system, the proof actually applies in any infinite number system.

## Equations of the First Degree

We consider the equation

$$ax + b = 0,$$

where $a$ and $b$ are rational numbers and $a$ is not zero. We claim that it has one and only one solution, namely $- b/a$. To see that $- b/a$ is actually a solution, we substitute $- b/a$ for $x$ in the left member, obtaining

$$a(- b/a) + b;$$

this equals

$$- b + b = 0.$$

To show that $- b/a$ is the *only* solution, we argue as follows. If

$$ax + b = 0$$

then

$$(ax + b) - b = 0 - b = - b.$$

Now

$$(ax + b) - b = ax + (b - b) = ax + 0 = ax,$$

so that

$$ax = - b.$$

Thus, if $ax + b = 0$ then $ax = -b$. Furthermore, from

$$ax = -b$$

we have

$$(1/a)(ax) = (1/a)(-b) = -b/a$$

and

$$(1/a)(ax) = ((1/a)(a))x = 1 \cdot x = x,$$

whence

$$x = -b/a.$$

Thus, if $ax + b = 0$ and $a$ is not zero, then $x = -b/a$.

*Example*: We consider a puzzle which can be rephrased as a problem that requires the solution of a first degree equation. "Mary is 24 years old. She is twice as old as Ann was when Mary was as old as Ann is now. How old is Ann?"

*Solution*: We designate the present age of Ann by $x$. We know that at a certain time in the past Mary's age was $x$ and Ann's age was 12. From that time to the present, Mary's age acquired $24 - x$ years and Ann's acquired $x - 12$. Since they age at the same rate, we have $24 - x = x - 12$, or $36 - 2x = 0$, from which we have $x = 18$.

## Quadratic Equations

We consider the equation $ax^2 + bx + c = 0$, where $a$, $b$, $c$ are rational numbers and $a$ is not zero. Such an equation is known as a *quadratic* equation. We describe a procedure for finding all the solutions of this equation. We first consider a special type of quadratic equation, namely $x^2 + c = 0$. This is special because the coefficient of $x$ is chosen to be zero and that of $x^2$ to be one. We add $-c$ to both sides and obtain

$$(1) \qquad\qquad x^2 = -c.$$

We note that any solution of this equation must be a solution of the original equation, and vice versa. To solve (1), we must find a rational whose square is $-c$. This is clearly impossible if $-c$ is negative. If $-c$ is zero, then $x^2 = 0$, and zero is the only solution. If $-c$ is not zero and is the square of some rational $w$, then both $w$ and $-w$ are solutions. They are the only solutions, because the equation, being of degree two, can have at most two solutions. To sum up: For the equation to have a solution $-c$ must be positive, but this is not enough; $-c$ must be, in addition, the square of a rational.

*Examples:* The equation $x^2 - 1 = 0$ is rewritten as

$$x^2 = 1,$$

from which we see that 1 and $-1$ are solutions (we have $c = -1$, $-c = 1 = 1^2 = (-1)^2$).

After the equation $x^2 - 2 = 0$ is rewritten as $x^2 = 2$, the question arises of the existence of a rational number whose square is two. The answer is not obvious. It can be shown that there is no such rational number. We leave the question open here, and give a proof in Chapter XI.

We now return to our general quadratic $ax^2 + bx + c = 0$ and show how to base its treatment on the case $x^2 + c = 0$. The crucial detail of the procedure depends on a useful device known as *completing the square*. Let $t$ be a rational number, and consider the expression $x^2 + tx$. If we add $t^2/4$, we obtain the new expression $x^2 + tx + t^2/4$ which is equal to the expression $(x + t/2)^2$, and this is obviously the square of an expression. For example, if we begin with $x^2 + 6x$ and add 9 to it (here $t = 6, t^2/4 = 9$), we obtain $x^2 + 6x + 9$, which equals $(x + 3)^2$.

Let us return to $ax^2 + bx + c = 0$. If we multiply both sides by $1/a$, we obtain

$$x^2 + (b/a)x + c/a = 0.$$

If we then add $-c/a$ to both sides, we obtain

$$x^2 + (b/a)x = -c/a.$$

We now use the device of completing the square and identify the number $b/a$ in the left member with the number $t$ of our above discussion of that subject. Since we added $t^2/4$ earlier, we now add $(b/a)^2/4 = b^2/4a^2$ and do this to both sides to obtain the new equation

$$x^2 + (b/a)x + b^2/4a^2 = b^2/4a^2 - c/a,$$

or

$$(x + b/2a)^2 = b^2/4a^2 - c/a.$$

All we need do now is consider $x + b/2a$ as a new variable $X$, the rational $b^2/4a^2 - c/a$ as a new rational $-C$, and proceed as in the earlier special case. We now distinguish two possibilities. If the rational number $b^2/4a^2 - c/a$ is not the square of some rational number, then for no rational number $x$ can $(x + b/2a)^2$, which is the square of a rational number, equal $b^2/4a^2 - c/a$. In this case, therefore, the original equation has no solutions. If, on the other hand, $b^2/4a^2 - c/a$ is the square of some rational number, the equation does have solutions. In fact, if $w$ is a rational for which

$$(2) \qquad\qquad w^2 = b^2/4a^2 - c/a,$$

then the solution of

$$x + b/2a = w,$$

namely $- b/2a + w$, is a solution of the original equation. This can be checked by substituting $- b/2a + w$ for $x$ in the original equation and replacing $w^2$ by its value as given above.

This procedure, in general, furnishes two solutions. For, if we can find a rational $w$ satisfying (2), its negative $-w$ is also a rational satisfying (2) and, unless $w = - w = 0$, the numbers $- b/2a + w$ and $- b/2a - w$ furnish two distinct solutions of the equation. Note that there can be no others, because Theorem III above states that there are at most two solutions.

*Examples*: Consider the equation

$$2x^2 + 3x - 5 = 0.$$

Our first step is to multiply by $1/2$, obtaining

$$x^2 + (3/2)x - 5/2 = 0.$$

Next, we add $5/2$ to both sides, obtaining

$$x^2 + (3/2)x = 5/2.$$

We then complete the square by adding 9/16 to both sides $(t = 3/2, t^2/4 = (9/4)(1/4) = 9/16)$, obtaining

$$x^2 + (3/2)x + 9/16 = 5/2 + 9/16$$

or

$$(x + 3/4)^2 = 49/16 = (7/4)^2.$$

Then we solve $x + 3/4 = 7/4$ and $x + 3/4 = -7/4$, obtaining 1 and $-5/2$ as solutions of the original equation.

Now consider the equation

$$2x^2 + 3x + 5 = 0.$$

Our procedure yields, in turn, the equations

$$x^2 + (3/2)x + 5/2 = 0,$$
$$x^2 + (3/2)x = -5/2,$$
$$x^2 + 3x/2 + 9/16 = -5/2 + 9/16,$$
$$(x + 3/4)^2 = -31/16.$$

We observe that $-31/16$ is not the square of a rational number, since it is negative, and we conclude that the original equation has no solutions.

Let us now consider a problem from the domain of physics which can be solved by means of quadratic equations. Suppose an object is thrown upward with a speed of $v$ feet/second, and we are required to find the number of seconds for it to attain a height of exactly 100 ft. If we call this number $x$, then under certain conditions, physical theory states that $x$ is a solution of the quadratic equation $-16x^2 + vx - 100 = 0$. Whether this equation has no solution, one solution, or two solutions depends on the value of $v$. This corresponds to what one might expect without using mathematics, for the object might never attain the height of 100 ft., or might just attain it and then fall, or attain it, go up farther, and pass the 100-ft. level again on the way down. While this example cannot be completely understood without some knowledge of physics, it does indicate that solving equations can be related to other activities.

## Equations with More than One Variable

We consider two polynomials $E$ and $F$, involving the variables $x$ and $y$, and study the equation $E = F$. We show that if $E$ and $F$ are in standard form they are not equal for all values of $x$ and $y$ unless they have precisely the same terms. By showing this we settle our problem of the equality of expressions involving two variables. If $E$ and $F$ are in standard form and do not contain precisely the same terms then $E - F$ contains a term $t x^p y^q$ with $t$ not zero. We can write

$$E - F = (a x^m + b x^{m-1} + \ldots + g) y^n + \text{ other terms } x^r y^s$$

where the exponent of $y$ in the other terms is less than $n$. We are going to find a value for $x$ and a value for $y$ for which $E - F$ does not have the value zero, and thus for which $E$ and $F$ do not have the same value. To this end, choose any value for $x$ which is not a solution of $a x^m + b x^{m-1} + \ldots + g = 0$. This is possible if our number system has infinitely many members, since the equation has at most $m$ solutions. Substitute this value for $x$ in $E - F$, obtaining

$$k y^n + \text{ other terms such as } l y^s,$$

with $s$ less than $n$ and with $k$ a non-zero member of the number system. There are at most $n$ values for $y$ that give this expression the value zero, so there must exist a number for which it does not have the value zero. This is our desired result. We can follow a similar procedure for expressions involving any number of variables. We omit the details.

An equation in several variables—unlike an equation in one variable—has in general an infinite number of solutions no matter what its degree. We illustrate this with an example of the first degree: $5x - 9y - 160 = 0$. To solve this equation, choose any value $q$ whatsoever for $x$ and substitute this value for $x$. What results is $5q - 9y - 160 = 0$, an equation involving one variable $y$, which can readily be solved. Thus for each value of $x$ there is one and only one value of $y$ which together with that $x$ furnishes a solution of the equation. It might be of interest to observe that this equation

$5x - 9y - 160 = 0$ describes the relation between the Fahrenheit and the Centigrade temperature scales. In fact, if we select a value for $x$ as above and identify it as the Fahrenheit temperature reading of an object, the value of $y$ determined by solving the equation will give the corresponding Centigrade temperature reading. The fact that this equation has infinitely many solutions but that not all pairs of numbers are solutions corresponds to the physical fact that infinitely many temperatures are possible but that for any given Fahrenheit reading the Centigrade reading is determined uniquely.

## Simultaneous Equations

Given several equations all involving the same variables, we can sometimes find a set of values of the variables that satisfy all the equations simultaneously. We investigate this state of affairs. We indicate that we seek common solutions to all of a set of equations by listing the equations one under the other as

$$E = F,$$
$$G = H,$$
$$\cdots\cdots\cdots$$
$$R = T.$$

Such a list is called a *system of simultaneous equations*. A single set of values for the variables which is a solution of each of the individual equations in the system will be called a *solution* of a system of simultaneous equations. We shall study in detail only the simplest system

$$ax + by = c,$$
$$dx + ey = f,$$

where $a$, $b$, $c$, $d$, $e$, and $f$ are numbers, and $x$ and $y$ are variables. We first comment briefly on the cases where more than two equations or more than two variables are involved or where the degree is greater than one. Perhaps the most striking fact in the study of these systems is the analogy they offer to Theorem III, namely, if there are $n$ variables and $n$ equations of degrees $a, b, \ldots, q$, respectively, then there are in general no more than $a \cdot b \cdot \ldots \cdot q$ solutions. Note that this is consistent with Theorem III, where there was one equation and one variable. The generalization of Theorem III just mentioned has certain exceptions, and a major undertaking of higher mathematics has been to examine these exceptions and to explore their consequences.

We use a highly diluted version of this investigation in our study of the system

$$ax + by = c,$$
$$dx + ey = f.$$

We proceed by multiplying both sides of the first equation by $d$ and both sides of the second by $a$, obtaining

$$dax + dby = dc,$$
$$adx + aey = af.$$

We then subtract the second equation from the first, obtaining

$$dby - aey = dc - af$$

or

$$(db - ae)y = dc - af.$$

If $db - ae$ is not zero, we find that

$$y = (dc - af)/(db - ae).$$

By multiplying the original equations by $e$ and $b$ and subtracting the second from the first, we obtain, similarly,

$$x = (bf - ec)/(db - ae).$$

Thus, if $db - ae$ is not zero we obtain exactly one solution of the system; this is consistent with the estimate of the number of solutions mentioned earlier. If $db - ae$ is zero, the method fails. It turns out that under this circumstance there may be no solutions or there may be infinitely many solutions. We shall not treat this case.

*Examples*: Consider the system

$$2x - 3y = 1,$$
$$3x + 4y = 10.$$

Following the procedure, we write

| $6x - 9y = 3$ | (multiplying by 3) |
| $6x + 8y = 20$ | (multiplying by 2) |
| $-17y = -17$ | (subtracting), |

whence $y = 1$. Similarly, we find $x = 2$. We now present a puzzle whose solution can be effected by solving a system of two simultaneous equations of this sort.

*Puzzle*: When rowing in the direction of the current of a river, a man can travel 6 miles in an hour, whereas when rowing against the current he only travels 3/2 miles in an hour. How fast does he row, and how fast is the current?

*Solution*: Let $x$ be the number of miles rowed in an hour in still water, and let $y$ be the speed in miles per hour of the current. Then

$$x + y = 6$$
$$x - y = 3/2,$$

and by applying the procedure just described, we find

$$x = 15/4, y = 9/4.$$

## Additional Comments

The theory of algebraic equations is a vast one, and this chapter has only treated a small portion of it. We now sketch some additional facts.

For the case of one variable, equations of degree three and four can be handled by procedures analogous to those we used for quadratic equations. It has been proved that no such procedures exist for equations of higher degree. It has also been proved that an equation of degree $n$ over any number system always has solutions provided one is willing to extend the number system and to accept as a solution a member of the extended system; also, that there are always exactly $n$ solutions provided one counts them suitably. Moreover, there is an important number system, the complex number system, containing all the major number systems we construct in this text, which has the property that equations formed over it have all their solutions in it and no extension is necessary.

In the case of several variables, the solution of systems of simultaneous first-degree equations in any number of variables is completely settled, and is not drastically different from our treatment of the two-variable case. The general theory, without restriction to the first degree, has been treated very successfully but is still the subject of research.

We also venture a comment on the practical side of the subject. It has been stated that some questions arising in other sciences can be answered by solving equations. To some extent this is due to the fact that in algebra one deals with the structure of number systems rather than with the properties of particular numbers. Thus even if one doesn't know a certain number but only knows some relations it bears to other numbers, one can still deploy the whole apparatus of algebra toward its discovery.

## EXERCISES

1. Prove that there is a polynomial $Q(x)$ which is such that $x^3 - x^2 - x - 2 = (x - 2) Q(x)$. Use Theorem I but not Theorem II.

2. Solve for $x$, where possible:

   $x + 3 = 0$. $\qquad$ $3x + 4 = 3x + 5$.

   $2x + 3 = 0$. $\qquad$ $(1/3)x + 1/4 = (1/5)x + 1/6$.

   $2x - 3 = 0$. $\qquad$ $ax = b$.

   $3x + 4 = 5x + 6$. $\qquad$ $ax + 2x + 3 = 0$.

   $3x + 4 = 3x + 4$. $\qquad$ $ax + bx + c = 0$.

3. Solve for $x$, where possible:

$x^2 - 1 = 0.$            $x^2 + 2x - 1 = 0.$

$x^2 - 4 = 0.$            $x^2 + 2x + 2 = 0.$

$x^2 - 5 = 0.$            $3x^2 + 4x + 5 = 0.$

$x^2 + 2x = 0.$            $10x^2 - 7x + 1 = 0.$

$x^2 + 2x + 1 = 0.$            $x^2 - a = 0$

$\qquad\qquad\qquad\qquad\qquad$ ($a$ is a rational).

4. Solve for $x$ and $y$:

a) $x - y = 1,$            c) $2x - 3y = 4,$

$\quad x + y = 3.$            $\quad -5x + 6y = 7.$

b) $2x - y = 1,$            d) $(\tfrac12) x - (\tfrac13) y = \tfrac14,$

$\quad x + y = 3.$            $\quad (\tfrac15) x - (\tfrac16) y = \tfrac17.$

e) $px + qy = 6$

$\quad rx + sy = 7$ ($p$ and $q$ are rationals).

5. a) If $p$ and $q$ are solutions of the equation $x^2 + bx + c = 0$, prove that $p + q = -b$ and $pq = c$.

b) If $p$ and $q$ are the solutions of the equation $ax^2 + bx + c = 0$, where $a \neq 0$, prove that $p + q = -b/a$ and $pq = c/a$.

6. It is given that $b$ and $c$ are special rationals (that is, rationals of the form $p/1$, $q/1$, where $p$ and $q$ are integers). Prove that if the equation $x^2 + bx + c = 0$ has a rational solution then it has two solutions that are themselves special rationals.

7. Assuming that the number system is the rational number system, find all the solutions of each of the following:

a) $3x + 0y = 4,$            c) $3x - y = 4,$

$\quad 2x + 0y = 1.$            $\quad 6x - 2y = 7.$

b) $3x - y = 4,$            d) $3x - y = 4.$

$\quad 6x - 2y = 8.$

8. Find an equation for which the set of all solutions is $\{1,2,3,4\}$.

# CHAPTER X

## ORDER

### Introduction

THIS CHAPTER COVERS some of the mathematics that has been inspired by the everyday ideas of *greater* and *less*. While these ideas are no longer considered an essential part of the properties of numbers, they do figure in the analysis of many important number systems. The intent of these terms is incorporated into the mathematical structure by means of formal definitions and proofs. The reader is urged to distinguish between his everyday use of the words and the mathematical one; otherwise he will not be impressed by their similarities nor take seriously their differences. Most of these differences arise in the consideration of infinite sets and have little to do with everyday life.

### Order for the Cardinals

We define the concept of *greater* for pairs of cardinals.

DEFINITION: Let $a$ and $b$ be any cardinals. If $a$ and $b$ are distinct, either there is a cardinal $c$ different from zero and such that $a = b + c$ or else there is a cardinal $d$ different from zero and such that $b = a + d$. If $a = b + c$ with $c$ not zero, we shall say that $a$ is *greater than* $b$ and $b$ is *less than* $a$. Similarly, if $b = a + d$ with $d$ not zero, we shall say that $b$ is greater than $a$ and $a$ is less than $b$. Note that if we had both $a = b + c$ and $b = a + d$, we would have, by addition,

$$a + b = b + c + a + d = a + b + c + d.$$

It would follow that $c + d = 0$, so that both $c$ and $d$ would have to be zero, whence $a = b$. Thus, for any two cardinals $a$ and $b$, one and only one of the statements

$a$ is less than $b$,

$a$ equals $b$,

$a$ is greater than $b$,

must hold.

We use symbols $>$ and $<$ to indicate these relations, in the following way: $a > b$ means and is read, "$a$ is greater than $b$"; $a < b$ means and is read, "$a$ is less than than $b$."

Note that while there are infinitely many cardinals greater than $a$, namely $a + 1$, $a + 2$, $a + 3$, etc., there are only finitely many cardinals less than $a$.

We state and prove some theorems that express order properties of the cardinal number system.

THEOREM I: *If $a$, $b$, and $c$ are cardinals and if $a > b$ and if $b > c$, then $a > c$.*

*Proof*: Using the definition of *greater than*, we conclude from $a > b$ and $b > c$, that $a = b + u$, $b = c + v$, where $u$ and $v$ are non-zero cardinals. It follows that

$$a = (c + v) + u = c + (v + u).$$

Thus $a > c$, because we can obtain $a$ by adding to $c$ the non-zero cardinal $v + u$.

*Note*: This theorem can also be stated in terms of the relation $<$, as follows: If $c < b$ and $b < a$, then $c < a$.

*Note*: It follows directly from this theorem that any finite set of cardinals has a greatest and a least. A similar statement is true of any number system in which the counterpart of Theorem I holds. It is easy to see that every infinite set of cardinals must have a least member but not a greatest member.

THEOREM II: *If $a$, $b$, and $u$ are cardinals, and if $a > b$, then $a + u > b + u$.*

*Proof:* We rewrite $a > b$ as $a = b + c$, where $c$ is not zero. Then

$$a + u = (b + c) + u = (b + u) + c,$$

whence

$$a + u > b + u.$$

THEOREM III: *If $a$, $b$, $p$, and $q$ are cardinals for which $a > b$ and $p > q$, then $a + p > b + q$.*

*Proof:* We rewrite $a > b$, $p > q$ as $a = b + c$, $p = q + r$, where $c$ and $r$ are non-zero cardinals. By addition, we have

$$a + p = (b + c) + (q + r) = (b + q) + (c + r).$$

Then since $c + r$ is not zero, we have $a + p > b + q$.

THEOREM IV: *If $a$, $b$, and $u$ are cardinals for which $a > b$ and if $u$ is not zero, then $ua > ub$.*

*Proof:* We rewrite $a > b$ as $a = b + c$, where $c$ is a non-zero cardinal. Multiplying by $u$, we have $ua = u(b + c) = ub + uc$. Since $u$ is not zero and $c$ is not zero, neither is $uc$. Then $ua$ is the sum of $ub$ and the non-zero cardinal $uc$, whence $ua > ub$.

THEOREM V: *If $a$, $b$, $p$, and $q$ are cardinals for which $a > b$, $p > q$ then $ap > bq$.*

*Proof:* We rewrite $a > b$, $p > q$ as $a = b + c$, $p = q + r$ where $c$ and $r$ are non-zero cardinals. By multiplication we have $ap = (b + c)(q + r) = bq + (br + cq + cr)$, and since the cardinal in the last pair of parentheses is not zero, $ap > bq$.

## Order for the Integers

We define the concepts *greater* and *less* for the integers in terms of the concepts of positive and negative integers. Later we carry over this definition and its consequences to any number system that has positive and negative numbers. The reader should bear in mind that while every integer has a negative, not every integer is negative, and that $-a$, which denotes the negative of $a$, need not be negative. (If $a$ is $-5$, $-a$ is $5$.)

If $a$ is a positive integer, we shall say $a > 0$ (read: $a$ is greater than zero), and if $a$ is a negative integer, we shall say $a < 0$ (read: $a$ is less than zero). If $a$ and $b$ are any integers, we shall say $a > b$ ($a$ is greater than $b$) if $a - b > 0$, and if $a - b < 0$, we shall say $a < b$. Thus, for any pair of integers $a$, $b$ one and only one of the statements $a > b$, $a = b$, $a < b$ must hold, since the integer $a - b$ must be either positive, zero, or negative. It turns out that under this definition of *greater* the positive integers are related in exactly the same way as their counterparts in the cardinals. Thus we have $5 > 2$ both for cardinals ($5 = 2 + 3$) and for integers ($5 - 2 = 3$, which is positive). It also turns out that each positive integer is greater than each negative integer. In comparing two negative integers, we are committed by the definition to such statements as $-2 > -1000$ (because $-2 - (-1000) = -2 + 1000 = 998$, which is positive). We can now prove for integers the counterparts of Theorems I-IV of the last section.

THEOREM I (for integers): *If $a$, $b$, and $c$ are integers and if $a > b$ and $b > c$, then $a > c$.*

*Proof:* $a > b$ implies that $a - b$ is positive, and $b > c$ implies that $b - c$ is positive. From $a - c = (a - b) + (b - c)$, we see that $a - c$ is the sum of positive integers and is therefore positive. Then, by definition, $a > c$.

THEOREM II (for integers): *If $a$, $b$, and $u$ are integers, and if $a > b$, then $a + u > b + u$.*

*Proof:* To show that $a + u > b + u$, we must show that $(a + u) - (b + u)$ is positive. But this expression equals $a - b$, and we know from $a > b$ that $a - b$ is positive.

THEOREM III (for integers): *If $a$, $b$, $p$, and $q$ are integers for which $a > b$ and $p > q$, then $a + p > b + q$.*

*Proof:* We are to show that $(a + p) - (b + q)$ is positive. This expression equals $(a - b) + (p - q)$, which is positive because it is the sum of two integers $a - b$ and $p - q$ that are both positive (since $a > b$ and $p > q$).

THEOREM IV (for integers): *If $a$, $b$, $u$ are integers for which $a > b$, $u > 0$, then $ua > ub$.*

*Proof*: We are to show that $ua - ub$ is positive. This expression equals $u(a - b)$ which is positive because $u$ is positive, $a - b$ is positive, and the product of positive integers is positive.

Thus the integers and the cardinals have much in common with respect to order. There are also some differences. For instance, not every set of integers has a least integer. Another difference is that there are infinitely many integers less than a given integer. Another is that the counterpart of Theorem V does not hold for integers, as the following example shows. We have

$$2 > -3, \; -4 > -5,$$

but

$$2 \cdot (-4) = -8,$$

which is not greater than

$$(-3)(-5) = 15.$$

Moreover, for integers we have a theorem that cannot even be stated for the cardinals.

THEOREM VI: *If a and b are integers, and $a > b$, then*

$$-a < -b.$$

*Proof*: We are to prove that $-a - (-b)$ is negative. This expression equals $b - a$ or $-(a - b)$. Since $a - b$ is positive $(a > b)$, $-(a - b)$ must be negative, which proves our statement.

### Order in General

We can now extend our study of order to number systems in general. We shall consider number systems that have a zero element and a one-element, in which each number has a negative and in which the product of non-zero numbers is itself a non-zero number. Let $N$ be any such number system. We shall show that if $N$ can be thought of as having positive and negative numbers, then $N$ has many other order properties like those of the system of integers. Specifically, we assume that

(1) for each number $a$ of $N$, one and only one of the following holds: $a$ is zero, $a$ is positive (written: $a > 0$), or the negative of $a$ is positive (written: $-a > 0$ or $a < 0$);

(2) the sum of two positive numbers is a positive number, and the product of two positive numbers is a positive number.

We note that the rules which decide which products are positive and which negative are the same for $N$ as for the integers. We have assumed that the product of positive numbers is positive. The product of two negative numbers must then be positive; this follows from the equation $ab = (-a)(-b)$, because if $a$ and $b$ are negative then $-a$ and $-b$ are positive, so that $(-a)(-b)$ is positive. Thus the square of any non-zero number is positive, and so is any sum of such squares. In particular, the one-element 1 is positive, because $1 = 1^2$. If $a$ is positive and $b$ is negative, then $ab$ is negative; this is because $-b$ is positive, whence $a(-b)$, or $-ab$, must be positive.

For any pair $a$, $b$ of members of $N$, we define $a > b$ ($a$ is greater than $b$) by the condition $a - b > 0$. Thus for any such pair, one and only one of the statements $a > b$, $a = b$, $a < b$ must hold. If $a$ is greater than $b$, we shall say $b$ is less than $a$ (written: $b < a$). We assert that Theorems I–IV and VI hold for $N$. We give no proofs here, since the proofs given for the case of integers apply verbatim to the present case.

We now introduce the terms *upper bound* and *lower bound*, which are basic in our further study of order and which are applicable in any ordered number system such as $N$.

DEFINITION: Let $S$ be a subset of $N$. If there is a number $b$ of $N$ such that no member of $S$ is greater than $b$, we shall say that $b$ is an *upper bound* for $S$. Similarly, if there is a number $c$ of $N$ such that no member of $S$ is less than $c$, we shall say that $c$ is a *lower bound* for $S$. If $S$ has an upper bound and a lower bound, we shall say that $S$ is *bounded*.

*Examples* (with integers): Let $S$ be the set $\{-5, 6, 11\}$. $S$ has upper bounds (11, or any number greater than 11); $S$ has lower bounds ($-5$, or any number less than $-5$); and $S$ is bounded.

The set of positive integers has lower bounds (1, 0, or any negative number) but no upper bounds.

The set $\{-2, -4, -6, \ldots\}$ has upper bounds ($-2$, or any number greater than $-2$) but no lower bound.

These concepts provide means for distinguishing different number systems. For instance, every set of cardinals has a lower bound, but this is not true for the integers. On the other hand, any bounded set of cardinals or of integers must have only a finite number of members. We shall see that this is not true of the rationals, when we discuss order for that system. An important order property of all the major number systems discussed in this book is known as the *Archimedean* property: For no positive number $a$ does the set $\{a, 2a, 3a, \ldots\}$ have an upper bound. To illustrate this property we use as an example a version of the observation of Archimedes, which caused the property to be named after him. Let $s$ denote the number of grains of sand on a beach, and let $a$ denote the number of grains of sand that can be carried in a cup. The observation is, that it is possible to excavate the beach completely with cups. In our language, if none of 1 cupful, 2 cupfuls, 3 cupfuls, etc. sufficed, then $s$ would be an upper bound for the set $\{a, 2a, 3a, \ldots\}$. Now if we believe that there is some cardinal number $s$ which gives the number of grains of sand on a beach, and if we know that the system of cardinal numbers has the Archimedean property, then $s$ cannot be thought of as an upper bound for the set $\{a, 2a, 3a, \ldots\}$. Consequently we would conclude that there is a member of the set, say $na$, greater than $s$, from which it would follow that at most $n$ cupfuls would excavate the beach.

There are ordered number systems that do not have the Archimedean property. In such number systems, if $b$ is an upper bound for the set of positive numbers $\{a, 2a, 3a, \ldots\}$ then $a$ is said to be *infinitesimal* with respect to $b$. Such number systems are quite esoteric, however; while the reader may encounter the term in old-fashioned calculus textbooks, calculus is based on the real number system, which has no infinitesimals.

Another aspect of the Archimedean property is its relation to the *long division* of elementary school. In long division one starts with two numbers, the dividend and the divisor, and finds a quotient and a remainder (all of them positive integers). In symbols: $p$ being the dividend and $q$ the divisor, long division yields an integer $a$, called the quotient, and an integer $b$, called the remainder, for which $p = aq + b$, where $b$ is less than $q$. In elementary school much stress is laid on the actual procedure for finding these two numbers. What we can do now is show that they exist in any number system having the Archimedean property. In fact, let $p$ and $q$ be any positive numbers, and consider the set of positive numbers $q$, $2q$, $3q$, etc.. Since $p$ is not an upper bound for these numbers, there is some positive integer $w$ for which $p$ is less than $wq$. Now the set of all such $w$'s must have a least. Call this least $a + 1$. Then $p$ is not less than $aq$, and $p$ is less than $(a + 1)q$. It follows that $p - aq$ is not negative and is less than $q$. Since $p = aq + (p - aq)$ we have, replacing $p - aq$ by the symbol $b$, that $p = aq + b$, where $b$ is non-negative and less than $q$.

## Order for the Rationals

We first show that our conclusions about order in number systems in general apply to the rational number system. Each rational number has a representation $a/b$, where $b$ is a positive integer and $a$ is an integer which may be positive, negative, or zero. We say that $a/b$ is *positive* if $a$ is positive, *zero* if $a$ is zero, and *negative* if $a$ is negative. Clearly, stipulations (1) and (2) of the preceding section hold, so that the conclusions based on them also hold.

The rest of our discussion of the rational number system from the standpoint of order is directed not so much to examining its resemblances to other number systems as it is to revealing its special properties. We know that the rational number system differs from the cardinals and the integers in that each of its non-zero members has a reciprocal. We shall show that this property together with the order properties distinguishes the rational number system from all other number systems. More precisely, we show that if $N$ is a number system of the type discussed in the preceding section, and if each non-zero member of $N$ has a reciprocal, then either $N$ is the rational number system or $N$ is an extension of the rational number system. Thus the rationals can be described as the "smallest" ordered number system in which non-zero members have reciprocals. This fact is supplemented in the next chapter by the introduction of the real number system, which is shown to be the "largest" such number system that has the Archimedean property.

To prove our claim about the rationals, we select the one-element of $N$, call it 1, and construct the numbers

$$1 + 1 = 2, \; 2 + 1 = 3, \; 3 + 1 = 4, \; \cdots.$$

Since 1 is positive, each of these numbers is positive, and since each is greater than its predecessor, these numbers are all different. (We have seen earlier that there are finite number systems; indeed, there is one in which $1 + 1 = 0$. Finiteness is excluded here by our order assumption.) Thus $N$ must contain $0, 1, 2, 3, \cdots$. Since $N$ must contain the negative of each of its members, it must also contain $-1, -2, -3, \cdots$. Since $N$ must contain the reciprocal of each of its non-zero members, it must also contain

$$1/2, \; 1/3, \; \ldots, \; -1/2, \; -1/3, \cdots.$$

Finally, since $N$ must contain all products of pairs of its members, it must contain all the rationals, which is what we set out to prove.

An order property of the rationals that distinguishes it from the cardinals and integers can be described by saying that between any two rationals there is a third. More precisely:

*If $a$ and $b$ are any distinct rationals and if $a < b$ then there is a rational $c$ which is such that $a < c < b$.*

Clearly, neither the cardinals nor the integers has this property. To prove the statement, we show that the rational $(a + b)/2$ has the desired property. To show that $(a + b)/2 < b$, we must show that $b - (a + b)/2 = (b - a)/2$ is positive. We know that $b - a$ is positive and that $1/2$ is positive; whence $(b - a)/2 = (1/2)(b - a)$ is also positive. To show that $a < (a + b)/2$ we must show that $(a + b)/2 - a = (b - a)/2$ is positive, and this has just been done.

It follows that between any two rationals there are infinitely many, because not only is there a rational $c$ between $a$ and $b$, but also a rational $d$ between $c$ and $b$ (and hence, by Theorem I, between $a$ and $b$), etc.

We now turn to the terms *greatest* and *least* and discuss some properties of the rational number system connected with them. In the first place, not every set of rationals has a greatest or a least, even if the set is bounded. Consider, for instance, the set of rationals greater than 0 and less than 1. It can have no greatest, because if $a$ is a member of the set, there is a rational between $a$ and 1, hence greater than $a$ and also in the set. Similarly, for each member of the set, there is a smaller number also in the set. Although this set has no greatest and no least member, it has in 1 and 0 something very much like a greatest and a least. Of course, neither 0 nor 1 is in the set, because the set was defined to exclude them. The relation of 1 to the set is that it is an upper bound and in fact the least upper bound; and 0 is, similarly, the greatest lower bound of the set. It is the ideas of least upper bound and greatest lower bound which furnish the mathematical equivalent, for infinite sets, of the ideas of least and greatest. Also, it is in terms of these ideas that a claim of insufficiency can be brought against the rational number system. In our next chapter, we shall see that there are bounded sets of rationals which have no least upper bound. In that chapter we construct the real number system, which is an extension of the rational number system, and which has the property that each of its bounded subsets has a least upper bound.

## EXERCISES

1. Let $N$ be the set of all polynomials in $x$ whose coefficients are integers. Define a member of $N$ to be positive if the coefficient of its highest power of $x$ is positive. Prove that

   a) $N$ is a number system relative to ordinary addition and multiplication.
   b) $N$ is ordered.
   c) $x$ is an upper bound for $\{1, 2, 3, \cdots\}$.
   d) $x^2$ is an upper bound for $\{x, 2x, 3x, \cdots\}$.

2. Given that $a$ and $b$ are members of an ordered number system and that $a > b$. Which of the following can be proved from this information?

$1/b > 1/a$, $-b > -a$, $a^2 > b^2$, $a^2 > ab$, $ab > b^2$, $2a > a + b$, $a^3 > a^2 b$, $a^2 + b^2 > 2ab$.

For those which cannot be proved, find additional hypotheses under which they can be proved and prove them under these additional hypotheses. Try to make each statement as general as possible.

3. Construct a bounded infinite set of rationals that has
   a) a greatest and a least;
   b) a greatest but no least;
   c) a least but no greatest.

4. Show that the set of positive rationals whose square is less than 10 has an upper bound. Show that the set of upper bounds for this set has no upper bound but does have a lower bound.

5. Show that
   a) $1 > -2$;
   b) $-3 > -4$;
   c) $4 > 3$;
   d) $0 > -8$;
   e) zero is greater than every negative integer;
   f) every positive integer is greater than every negative integer;
   g) every positive integer is greater than zero.

6. If $b$ is a rational,
   a) is $-b$ a negative number?
   b) is $-b^2$ a negative number?
   c) is $(-b)^2$ a negative number?

7. Show that in any ordered number system
   a) If $a > 0$ and $b < 0$, then $ab < 0$.
   b) If $a > 0$ and $1/a$ exists, then $1/a > 0$.

8. Let $C$ be the set of all cardinal numbers. We now define two new operations on $C$. If $a$ and $b$ are distinct cardinal numbers, we define as their "max" and designate by the symbol $a \vee b$ the greater of the two numbers $a$ and $b$. We complete the definition of this operation by specifying that $a \vee a$ is $a$. Similarly, we define as the "min" of $a$ and $b$ and designate by the symbol $a \wedge b$ the lesser of the two numbers $a$ and $b$. Again, we complete the definition by specifying that $a \wedge a$ is $a$.

a) Show that these operations are associative and commutative.

b) Show that only one of them has an identity element.

c) Show that $+$ is distributive with respect to $\vee$. Thus we have a number system in which the multiplication operation is $+$ and the addition operation is $\vee$. What is its one-element? its zero-element?

# CHAPTER XI

## THE REAL NUMBER SYSTEM

### Introduction

In this chapter we construct the real number system and examine some of its properties. We take up the extraction of roots, general exponents, and logarithms, concentrating as usual on their nature as mathematical objects rather than on their manipulation. In some cases proofs have been omitted, but in each case the omitted proof can be made up from material given in the text, along lines similar to proofs actually used.

A single simple fact, formulated as Theorem I of this chapter, is crucial in the working out of these matters. Actually, this theorem does more than provide a basis for developing this portion of mathematics. It provides a means of throwing light on many obscurities, from Zeno's paradoxes down to quite recent philosophical speculations. Much has been written about continuous magnitudes, both in and out of mathematical contexts, and this includes a great deal of nonsense. The clarity that results from approaching some of these problems along lines suggested by Theorem I is astonishing.

### Finite Decimals

Rational numbers whose denominators are powers of 10 are often designated by decimals. Thus the rational $1576/10^2$ is written as $15.76$, the rational $1576/10^5$ as $.01576$. By a *finite decimal* we mean any symbol of the form

$$I . a_1\, a_2 \cdots a_r$$

or of the form

$$- I . a_1\, a_2 \cdots a_r$$

where $I$ is any non-negative integer, and $a_1, a_2, \ldots, a_r$ are non-negative integers less than 10. We refer to $I$ (and $-I$, respectively) as the *integral part* of the decimal. Such a symbol will be taken to designate the rational number

$$I + a_1(1/10) + a_2(1/10)^2 + \cdots + a_r(1/10)^r,$$

and the rational number

$$- I - a_1(1/10) - a_2(1/10)^2 - \cdots - a_r(1/10)^r$$

respectively. Each of these can evidently be written as a single fraction whose denominator is a power of 10.

The set of all such rational numbers constitutes a number system which has a zero-element 0., a one-element 1., and in which each number has a negative. It is instructive to see that the familiar rules for adding and multiplying finite decimals give results completely consistent with the results of these operations when applied to their fractional counterparts. The interested reader can verify this without difficulty.

This number system does not have the property of containing the reciprocal of each of its non-zero members. For instance, the rational number 3. is a finite decimal, but while 3. has a perfectly good reciprocal 1/3, this reciprocal is not a finite decimal. This is not too serious a defect for practical computation because we can find, in the finite decimals .3, .33, .333, etc., arbitrarily good approximations to 1/3.

Since these finite decimals represent rationals, they are ordered. It is a useful fact that given two positive finite decimals, it can be determined without calculation which is greater, just as with integers. More specifically, if

$$a = I.a_1 a_2 \cdots a_r$$

and

$$b = J.b_1 b_2 \cdots b_s$$

are any positive finite decimals, our test for $a > b$ is that $a - b > 0$. The outcome of this test can be predicted as follows. If $I$ and $J$ are unequal, then $a > b$ if and only if $I > J$. If $I = J$ but $a_1$ and $b_1$ are not equal, then $a > b$ if and only if $a_1 > b_1$. In general, we examine the first place in which the entries in $a$ and $b$ are different; the relative size of these entries determines the relative size of $a$ and $b$. Under certain circumstances the test as outlined is not decisive. For instance, in comparing $a = 1.3$ with $b = 1.34$, we find nothing in $a$ with which to compare the 4 of $b$. The solution is to adjoin enough zeros to make possible a decisive comparison. Thus, $1.3 = 1.30$; and for $a$ in this form the test works.

This kind of order is known as lexicographic order. It is named after the scheme for ordering words in a dictionary. Here our "words" are decimals, our first "letters" are any non-negative integers, and all subsequent "letters" are non-negative integers less than 10.

## The Rationals as Infinite Decimals

We are going to find a decimal representation for rational numbers. We know in advance that the system of finite decimals is insufficient for this purpose, and we turn to the set of infinite decimals. By an *infinite decimal* we mean a symbol $I.a_1 a_2 \cdots$ or $-I.a_1 a_2 \ldots$, in which $I$ is any non-negative integer and $a_1$, $a_2$, etc. are integers between 0 and 9 inclusive, and where such an $a_n$ is specified for each positive integer $n$. Our scheme for representing rational numbers by infinite decimals is an elaboration of the long-division process taught in elementary school. Since that process, though compact and efficient, is somewhat cryptic, the analysis of it that follows may incidentally help the reader to understand the mathematical ideas in back of it.

We begin with a numerical example. The fraction $70/19$ gives rise to the following array of calculations, which we assume will be familiar to the reader.

$$
\begin{array}{r}
3.684 \\
19 \overline{)70.000} \\
57 \\
\hline
13\ 0 \\
11\ 4 \\
\hline
1\ 60 \\
1\ 52 \\
\hline
80 \\
76 \\
\hline
4
\end{array}
$$

We shall return to this array after describing the procedure for obtaining an infinite decimal from any rational number. It is convenient to deal with positive rationals. The case of a negative rational can be treated by first handling its negative (which is a positive number), and then putting a minus sign in front of the result.

Let $m$ and $n$ be any positive integers. We find the digits for our representation of $m/n$ as an infinite decimal by repeated use of the division process described in the discussion of the Archimedean property. This process enables us to find pairs of non-negative integers $(I, p)$, $(a_1, p_1)$, $(a_2, p_2)$, $(a_3, p_3)$, $\cdots$, for which

$$
\begin{aligned}
m &= I n + p \\
10p &= a_1 n + p_1 \\
10p_1 &= a_2 n + p_2 \\
10p_2 &= a_3 n + p_3
\end{aligned}
$$

$$\cdots\cdots\cdots\cdots\cdots$$

where each of the remainders $p$, $p_1$, $p_2$, $p_3$, $\cdots$, is less than $n$. We further claim that the quotients $a_1$, $a_2$, $a_3$, $\cdots$ are each less than 10; if, for instance, $a_2$ were as great as 10, then $10p_1$, being equal to $a_2 n + p_2$, would certainly be as great as $10n$. This is not consistent with the fact that $p_1$ is less than $n$. It follows that the integers $I, a_1, a_2, a_3, \cdots$ are acceptable ingredients for an infinite decimal $I.a_1 a_2 a_3 \cdots$. This is the desired representation for $m/n$.

In the case of the particular rational number $70/19$, the corresponding first four steps are

$$70 = (3)(19) + 13$$
$$130 = (6)(19) + 16$$
$$160 = (8)(19) + 8$$
$$80 = (4)(19) + 4$$

and the corresponding infinite decimal representation is $3.684 \cdots$. Let us now examine the connection between rational numbers and their decimal representations. We deal with the numerical example first. If we replace the 13 in the first equation by its value

$$(6/10)(19) + 16/10$$

as determined from the second equation we obtain

$$70 = (3)(19) + (6/10)(19) + 16/10$$
$$= (3.6)(19) + 16/10.$$

If we replace the $16/10$ of this equation by its value

$$(8/10^2)(19) + 8/10^2$$

as determined from the third equation we obtain

$$70 = (3)(19) + (6/10)(19) + (8/10^2)(19) + 8/10^2$$
$$= (3.68)(19) + 8/10^2.$$

If we replace $8/10^2$ by its value $(4/10^3)(19) + 4/10^3$ as obtained from the fourth equation, we obtain

$$70 = (3)(19) + (6/10)(19) + (8/10^2)(19)$$
$$+ (4/10^3)(19) + 4/10^3$$
$$= (3.684)(19) + 4/10^3.$$

By dividing by 19 we then have

$$70/19 = 3.6 + 16/190$$
$$= 3.68 + 8/1900$$
$$= 3.684 + 4/19000$$

The reader is invited to locate the places in the array given at the beginning of this section which effect the calculations just made.

To see the connection between $m/n$ and $I.a_1\,a_2\,a_3\cdots$ we perform similar calculations. We obtain successively

$$m = In + (a_1/10)n + p_1/10 = (I.a_1)n + p_1/10$$
$$m = (I.a_1)n + (a_2/10^2)n + p_2/10^2 = (I.a_1a_2)n + p_2/10^2$$
$$m = (I.a_1a_2)n + (a_3/10^3)n + p_3/10^3$$
$$= (I.a_1a_2a_3)n + p_3/10^3$$
$$\cdots\cdots\cdots\cdots\cdots\cdots$$

from which

$$m/n = I.a_1 + p_1/10n$$
$$m/n = I.a_1a_2 + p_2/100n$$
$$m/n = I.a_1a_2a_3 + p_3/1000n$$
$$\cdots\cdots\cdots\cdots\cdots\cdots\cdots$$

Therefore the finite segments $I.a_1,\ I.a_1\,a_2,\ I.a_1\,a_2\,a_3$ of the infinite decimal give better and better approximations to $m/n$ since the remainders $p_1/10n$, $p_2/100n$, $p_3/1000n$ are smaller and smaller. In the numerical example, 3.684 differs from 70/19 by only 4/19000, and arbitrarily close approximations can be obtained by carrying out a sufficiently great number of steps.

## Repeating Decimals

Two examples of repeating decimals are

$$.333\cdots$$

and $\qquad\qquad\qquad 14.68325325325\cdots.$

A decimal is said to be *repeating* if there is some block of digits $a\,b\cdots f$ which is such that from some point on the entries in the decimal are $a\,b\cdots f\,a\,b\cdots f\,a\,b\cdots f\cdots$. The reader has undoubtedly encountered such decimals; for instance, he has obtained $.333\cdots$ as a representation for 1/3.

We now show that every decimal that represents a rational number in the sense described in the previous section must be a repeating decimal. Recall that each of the remainders $p_1, p_2, \cdots$ is less than $n$. It follows that after at most $n-1$ steps a remainder must occur which was previously obtained. Suppose the remainder in question is $p$ and that the calculations which were performed till $p$ occurred again are

$$10p = an + q$$
$$10q = bn + r$$
$$\cdots\cdots\cdots\cdots$$
$$10t = fn + p.$$

Clearly, the next calculation must again yield $10p = an + q$, and the entries in the decimal must be thereafter $ab\cdots fab\cdots f\cdots$. This proves our assertion. This fact is not generally known empirically, even to people who perform many divisions. It is easy to see why; to obtain the repetition pattern with 70/19, for instance, may require as many as 18 digits, and hence 18 steps.

The converse of this proposition is also true, that each repeating decimal represents a rational. Let us see, for instance, how given the repeating decimal .333 $\cdots$ we could reconstruct the rational 1/3 which it represents. From $x = .333\cdots$ we obtain, by multiplying by 10,

$$10x = 3.33\cdots = 3 + .33\cdots = 3 + x,$$

whence $\qquad\qquad 9x = 3,\ x = 1/3.$

This procedure works for more elaborate cases, like 14.68325325 $\cdots$. In fact, let

$$x = 14.68325325\cdots = 14.68 + .00325325\cdots$$

and let

$$y = .325325\cdots.$$

Then

$$1000y = 325.325\cdots = 325 + .325\cdots = 325 + y$$

whence

$$999y = 325,\ y = 325/999.$$

Then

$$x = 14.68 + y/100,$$

which is clearly rational. It should be stated that our steps in these examples are not justifiable in terms of any of our previous work because they depend on operations with infinite decimals, and we have as yet no basis for performing such operations. Indeed, we have avoided up till now making the claim that a rational number equals an infinite decimal but have only said that it is represented by it. The steps are quite reliable, however, and they can be applied to any repeating decimal. The next sections furnish the necessary justification.

## Construction of the Real Number System

Our task of constructing the real number system consists in assembling a set whose members are to be the real numbers, defining two operations on the set, and showing that the operations obey the five fundamental laws. Our set is : the infinite decimal $.000 \cdots$ in which all the entries are zero, together with that set of infinite decimals $I.a_1 a_2 a_3 \cdots$ and $-I.a_1 a_2 a_3 \cdots$ in which infinitely many of the entries are digits other than zero. In other words, we take the set of all infinite decimals and remove those decimals that have after a certain point nothing but zeros. Each $a$, is of course, a non-negative integer less than 10. We call this set, the set $R$.

An operation on $R$ is the assignment of a member of $R$ to each ordered pair of members of $R$. We define two such operations. Let $p = I.a_1 a_2 a_3 \cdots$ and $q = J.b_1 b_2 b_3 \cdots$ be any two members of $R$. To find the entries for the infinite decimal which is to be $p + q$ we find successively $I + J$, $I.a_1 + J.b_1$, $I.a_1 a_2 + J.b_1 b_2$ $I.a_1 a_2 a_3 + J.b_1 b_2 b_3$, etc. Each of these sums of finite decimals is a finite decimal, and we claim that after some preliminary fluctuations the entry in any given place is stabilized as some definite digit. It is this digit which is to be used as the corresponding entry in $p + q$. For instance, if $p = 8.888 \cdots$ and $q = 1.21212 \cdots$, some of the successive sums are 9, 10.0, 10.09, 10.100, 10.1009, and $p + q$ is $10.101010 \cdots$. We define multiplication similarly by reference to the products

$$(I)(J), \ (I.a_1)(J.b_1), \ (I.a_1 a_2)(J.b_1 b_2),$$

etc.

We claim that these operations satisfy the five laws which certify $R$ as a number system, the *real number system*. The elements of $R$ will be called *real numbers*. We claim that $R$ has a zero-element, namely $0.000 \cdots$, and that $R$ has a one-element, namely

.999 ⋯. We claim further that each member of $R$ has a negative in $R$ and that each non-zero member of $R$ has a reciprocal in $R$. We give no proofs of these claims but assure the reader that proofs can be devised solely on the basis of earlier material of this chapter.

## Order Properties of the Real Number System

It is easy to see that the presence or absence of a minus sign in front of $I$ effects a partition of the non-zero members of $R$ into negative and positive numbers. That the positive reals meet the stipulations placed on positive numbers in Chapter X is an immediate consequence of the definitions of addition and multiplication for reals. In short, $R$ is an ordered number system. Indeed, the condition $a > b$ if $a - b$ is positive implies that the ordering of the members of $R$ is the lexicographic one for positive reals.

We show that $R$ has the Archimedean property. Let $p$ be any positive real, and consider the set of infinite decimals $p + p$, $p + p + p$, etc. Arbitrarily great integral parts can be found by going out sufficiently far in this set. Thus, no matter which real number $q$ we select, a sum $p + p + \cdots + p$ can be found whose integral part exceeds the integral part of $q$, so that $q$ is not an upper bound for the set.

A basic theorem in the theory of the real number system is the following:

THEOREM I: *Let $S$ be a set of real numbers, other than the empty set, which has an upper bound. Then there is a real number which is the least upper bound of $S$.*

We assume in our proof that $S$ contains some positive numbers. Minor modifications cover the contrary case.

*Proof*: Let $I^*$ be the greatest integer which is the integral part of some member of $S$. Such an $I^*$ must exist; otherwise the set of integral parts would have no upper bound and then $S$ would have no upper bound. Next, we select all the members of $S$ that have $I^*$ as integral part and examine the first digit after the decimal point in each. Let $a_1^*$ be the largest digit obtained in this way. We then select all the members of $S$ that begin with $I^*.a_1^*$ and examine the second digit after the decimal point in each. Let $a_2^*$ be the largest digit obtained in this way. We continue in this way and select $a_3^*, a_4^*, \cdots$. What we obtain is an infinite decimal $b^* = I^*.a_1^* a_2^* a_3^* \cdots$. This infinite decimal is a real number, because it cannot have all zero entries after a certain point.

We claim that $b^*$ is the least upper bound for $S$. To prove this, we must show that it is indeed an upper bound and that no smaller number is an upper bound. Let $p$ be any number greater than $b^*$. Then $p$ must be of the form $\cdots t \cdots$, where the dots before $t$ represent digits identical with the corresponding ones in $b^*$ and where $t$ is greater than the corresponding entry in $b^*$. Certainly $p$ is not in $S$ because in view of the construction of $b^*$, $t$ is too great. Thus no number $p$ greater than $b^*$ is in $S$, so that $b^*$ is an upper bound for $S$. Now let $q$ be any number less than $b^*$. Then $q$ must be of the form $\cdots s \cdots$ where the dots before $s$ represent digits identical with the corresponding ones in $b^*$ and where $s$ is less than the corresponding entry in $b^*$. Since there is in $S$ a member that has the same entries as $b^*$ up to and including the place occupied by $s$, there is in $S$ a member greater than $q$. Therefore no number $q$ less than $b^*$ is an upper bound for $S$ because given such a number $q$ we can find in $S$ a number greater than $q$. This shows that $b^*$ is the least upper bound for $S$.

A similar proof shows that if $S$ has a lower bound, it has a greatest lower bound.

Our next theorem shows that $R$ is the "largest" ordered number system of a certain kind.

THEOREM II: *Let $N$ be an ordered number system with the Archimedean property and which is an extension of the real number system. Let Theorem I hold in $N$. Then $N$ is the real number system.*

*Proof*: If $N$ were not identical with $R$, it would contain a number not in $R$. We may assume that this number is positive, for the negative of any number not in $R$ would also not be in $R$, and one of the two must be positive. (Zero is not available here, because zero is real.) Call this positive number $p$. Now the set of real numbers less than $p$ certainly has members: for instance, zero or any negative real. The set of real numbers greater than $p$ also has members, for the set consisting of the finite sums of the one-element,

all of whose members are reals, contains such numbers by virtue
of the Archimedean property. Call the first set $L$ and the second $G$.
Then $L$ has upper bounds (any member of $G$, for instance), and
$G$ has lower bounds (any member of $L$, for instance). Consequently, $L$ has a real least upper bound $l$ and $G$ has a real greatest
lower bound $g$. It is easy to see that $l < p < g$. Now consider the
real number $(l + g)/2$. It is greater than $l$ and less than $g$. Being
real and greater than $l$ it cannot be a member of $L$, because $l$ is an
upper bound for $L$. Being real and less than $g$ it cannot be a member
of $G$, because $g$ is a lower bound for $G$. Thus it is neither in $G$
nor in $L$, that is, it is neither greater than nor less than $p$. It must
therefore equal $p$. But this also is impossible, because $p$ is assumed
to be not real, and $(l + g)/2$ is real. Thus we are led to contradictions if we assume $N$ to be different from $R$. $N$ and $R$ must
therefore be identical, which is what we wanted to prove.

*Note*: It is possible to prove not only that $R$ is the "largest"
such number system, but that it is the only one. We sketch a proof
of this in the next section.

### The Reals as an Extension of the Rationals

We are going to show that the reals contain a subset isomorphic
to the rationals. To accomplish this, we must assign a real to each
rational and show that the reals assigned to sums and products of
pairs of rationals are the sums and products of the reals assigned
to the rationals in question. In making our assignments we use in
slightly modified form our process for representing rationals as
infinite decimals. The modification consists in choosing the remainders to lie between 1 and $n$ inclusive rather than between 0
and $n - 1$. This guarantees that zero never occurs as a remainder,
that is, that nothing "comes out even." It leads to somewhat bizarre
results for the fractions $p/1$ where $p$ is an integer: for instance,
from 1 (thought of as $1/1$ to institute the division) we obtain
$0.999 \cdots$, from 2 we obtain $1.999 \cdots$, etc. For rationals that do
not have representations as finite decimals, nothing is changed.
The modified process yields a repeating decimal in every case and
one, moreover, that is acceptable as a real number as well. We

assign to the rational $p/q$ the real number obtained from it by this modified division process. We omit the proof that these assignments constitute an isomorphism.

To show that the real number system contains numbers other than those assigned to rationals, all we need do is exhibit an infinite decimal that is not a repeating decimal. One such is $.1010010001\cdots$, where pairs of 1's enclose longer and longer strings of zeros. The existence of such numbers suggests an investigation of their utility. Our next sections are directed to this end.

We can now sketch the proof that the reals are the only number system of the type discussed in Theorem II. More precisely, we show that any ordered extension of the rationals in which Theorem I holds must contain all the reals. Since, by Theorem II, it contains no more than the reals, we see that it must be identical with the system of reals.

Let us first show that every real number is the least upper bound of some set of rationals. One way of doing this for the real number $I.a_1 a_2 \cdots$ is to take the set of rationals represented by the finite decimals $I.$, $I.a_1$, $I.a_1 a_2$, etc. Clearly their least upper bound is $I.a_1 a_2 \cdots$. Therefore any number system that contains the rationals and also all the least upper bounds of sets of rationals must contain the reals.

## Extraction of Roots

We return to the topic of equations and investigate the special equations $x^n = a$, where $a$ is a positive real number and $n$ is a positive integer. It turns out that each such equation has one and only one positive real solution. This solution is sometimes designated by the symbol $\sqrt[n]{a}$ (read: the $n$-th root of $a$) and sometimes by the symbol $a^{1/n}$ (read: $a$ to the $1/n$). The latter nomenclature will be discussed in the next section. We take up here the existence of the number in question.

We need certain facts about order among powers of numbers. One of them is that if $p$ and $a$ are positive reals, $n$ a positive integer and $p^n < a$, then there is a positive real $t$ which is such that $(p + t)^n < a$. Another is that, for similar $p$, $a$, and $n$, if $p^n > a$, then for some positive real $t$, $(p - t)^n > a$. A third is that if

$p$ and $q$ are positive reals and if $p > q$, then $p^n > q^n$ for every positive integer $n$. Let us prove the last of these statements. From $p > q$ we have $p = q + t$, where $t = p - q$ is a positive real number. Then

$$p^n = (q + t)^n = (q + t)(q + t) \cdots (q + t)$$
$$= q^n + \text{other positive terms.}$$

Thus $p^n$ is obtained from $q^n$ by adding certain positive numbers to it, whence $p^n > q^n$.

To prove the existence of a positive real solution of $x^n = a$ we consider the set $S$ of positive real numbers $s$ for which $s^n < a$. Such numbers exist: for instance, a high power of $(1/10)$ has so many zeros after the decimal point that it and its $n$-th power are surely less than $a$. We show that this set has an upper bound. Certainly for some power $u$ of 10, $(10^u)^n > a$. Then $10^u$ is an upper bound for $S$, because if $q > 10^u$, $q^n > (10^u)^n > a$ and therefore $q$ is not in $S$. By Theorem I, $S$ must have a least upper bound $b$. We claim that $b^n = a$. If this were not so, either $b^n < a$ or $b^n > a$. If $b^n > a$, there is a positive real $t$ for which $(b - t)^n > a$, whence $b - t$ is an upper bound for $S$ smaller than $b$, and this is inconsistent with the fact that $b$ is the least upper bound. If $b^n < a$, there is a positive real $t$ for which $(b + t)^n < a$, whence $b + t$ is in $S$ and is greater than $b$. This is inconsistent with the status of $b$ as an upper bound. Therefore, since $b^n > a$ and $b^n < a$ lead to contradictions, we must have $b^n = a$.

No such result holds in the rationals. To confirm this, we give Euclid's proof of the non-existence of a rational whose square is 2. If there were a rational whose square were 2, we should have a positive rational $p/q$, in lowest terms, which is such that $(p/q)^2 = 2$. From this we should have $p^2/q^2 = 2$ and then $p^2 = 2q^2$. We deduce from this that $p$ is even, say $2w$. Then $(2w)^2 = 2q^2$, $4w^2 = 2q^2$, $2w^2 = q^2$, whence $q$ is also even. Thus $p$ and $q$ have a common factor 2, which is inconsistent with the assumption that $p/q$ is in lowest terms. Since the assumption that such a rational exists leads to a contradiction, it follows that there is no such rational.

## Exponents

We first define exponents in general. We say that a number system $E$ is a *set of exponents* for a number system $N$ if there is assigned to each member $(n, e)$ of $N \times E$ a member $n^e$ of $N$, and if these assignments obey the rules

(1) $\qquad\qquad\qquad n^e \cdot n^f = n^{e+f}$

(2) $\qquad\qquad\qquad (n^e)^f = n^{ef}$

(3) $\qquad\qquad\qquad (nm)^e = n^e m^e$

and if $E$ has a one-element, $n^1 = n$. Clearly these rules are inspired by our results with the positive integers as exponents.

We examine the rational number system as a possible set of exponents for the positive reals. We show that there is only one way of making the necessary assignments, that is, that for each real $n$ and for each rational $e$ there is only one way of defining $n^e$ that is consistent with the above rules. We know that $n^1 = n$. Then $n \cdot n = n^1 \cdot n^1 = n^2$ (by rule (1)). Thus if we are to have exponents at all, $n^2$ must be $n \cdot n$. Similarly, $n^3$ must be $n \cdot n \cdot n$ and so on for all the positive integers. We now examine the possibilities for $n^{-1}$. From $n^{-1} \cdot n^2 = n^{-1+2} = n^1 = n$, and $n^{-1} \cdot n^2 = n^{-1}(n \cdot n) = (n^{-1} \cdot n)n$ we have $(n^{-1} \cdot n)n = n$, whence $n^{-1} \cdot n$ is the one-element. It follows that $n^{-1}$ must be the reciprocal of $n$. From $n^{-1} \cdot n = 1$ and $n^{-1} \cdot n = n^{-1} \cdot n^1 = n^{-1+1} = n^0$, we must have $n^0 = 1$. It is also possible to see what $n^{-t}$ must be, where $t$ is a positive integer. Rule (2) stipulates that $n^{-t} = (n^t)^{-1}$, which is the reciprocal of $n^t$. Let us summarize our results so far. For positive integral exponents $t$, $n^t$ must be the familiar product $n \cdot n \cdots n$, with $t$ factors. For negative integral exponents $-t$, $n^{-t}$ must be $(1/n)(1/n) \cdots (1/n)$ with $t$ factors. Finally, $n^0$ must be 1.

We turn now to fractional exponents. We consider first, for positive integral $t$, the exponent $1/t$. From $(n^{1/t})^t = n^1 = n$ we see that $n^{1/t}$ must be what we designated earlier by $\sqrt[t]{n}$. Then $n^{s/t} = (n^{1/t})^s$ must be the $s$-th power of the $t$-th root of $n$. We conclude that if any assignments that satisfy the above rules are at all possible, they must be these; no other set of assignments can be made. We have not shown, however, that this set of assignments actually has the desired properties. This can be done, but we omit the proofs.

Note that the use of negative integers as exponents requires only that the number system $N$ have reciprocals. Thus, the integers are a set of exponents for the rationals. On the other hand, fractional exponents require the extraction of roots, and this is not generally possible in the system of rationals.

## Logarithms

We give a brief sketch of what logarithms are and how they can be used. We require an even more complicated system of exponents than the rationals, namely the reals. If $n$ is a positive real and $e$ is a real, we take the set of rationals $p, q, r, \cdots$ for which $e$ was shown above to be the least upper bound and consider the set $n^p, n^q, n^r, \cdots$. If these numbers increase, we select their least upper bound; if they decrease, we select their greatest lower bound; and in either case call our selection $n^e$. The proof that these assignments actually make the reals a set of exponents for the positive reals is omitted.

By the *base* 10 *logarithm* of a positive real number $n$, we mean that real number $l$ for which $10^l = n$. More generally, by the *base b logarithm* of a positive real number $n$ we mean that real number $l$ for which $b^l = n$. If we have $10^l = n$, we write $l = \log_{10} n$, and for $a^l = n$, we write $l = \log_a n$.

Logarithms with base 10 and logarithms with certain other bases have been tabulated. By use of such tables it is possible to simplify certain calculations. For instance, the product $(4.4916)(4.5327)$ can be evaluated by logarithms as follows:

$$
\begin{aligned}
4.4916 &= 10^{.65240} & &\text{(from the tables)} \\
4.5327 &= 10^{.65636} & &\text{(from the tables)} \\
(4.4916)(4.5327) &= (10^{.65240})(10^{.65636}) \\
&= 10^{1.30876} & &\text{(by rule (1) for} \\
& & &\text{exponents)} \\
&= 20.359 & &\text{(from the tables).}
\end{aligned}
$$

The labor-saving character of logarithms is more strikingly revealed in the evaluation of $\sqrt[10]{7511}$. We have

$$^{10}\sqrt{7511} = (7511)^{1/10} \qquad \textbf{(by definition)}$$
$$= (10^{3.87570})^{1/10} \qquad \text{(from the tables)}$$
$$= 10^{.38757} \qquad \text{(by rule (2) for exponents)}$$
$$= 2.4410 \qquad \text{(from the tables)}.$$

*Note*: Calculations of this description only give approximations to the correct results. The tables supply a finite decimal approximation to the true logarithm, and the calculated results are finite decimal approximations to the correct ones. The relation of $^{10}\sqrt{7511}$ to 2.4410 is not truly one of equality but rather 2.4410 is the best finite decimal approximation to $^{10}\sqrt{7511}$ which uses that many digits.

It would be misleading to give the impression that logarithms are principally means for facilitating calculations. Indeed, nowadays calculating machines tend to replace the use of logarithms in this connection. It is as a theoretical instrument that logarithms find their chief role. In all the sciences, processes are studied which are described and understood in terms of logarithms. Many of the other mathematical objects examined in this book are also used in this way. And it is only when the mathematical nature of these objects is understood that they can be used to throw light on processes outside mathematics.

## EXERCISES

1. Express as the quotient of integers:
   $.222222\cdots$, $.252525\cdots$, $.00252525\cdots$, $1.2252525\cdots$.

2. Express as a repeating decimal:
   $1/2$, $1/3$, $1/9$, $1/7$.

3. Express as an infinite decimal:
   a) $(.111\cdots) + (.333\cdots)$;
   b) $(.111\cdots)(.333\cdots)$;
   c) $(.666\cdots) + (.777\cdots)$;
   d) $(.666\cdots)(.777\cdots)$.

4. Show that if $a$ is a positive real, there is a positive integer $n$ which is such that $1/n < a$.

5. Show that if $a$ is any positive real, there is a positive integer $n$ such that $1/2^n < a$.

6. Let $A$ and $B$ be non-empty sets of positive real numbers. Let $S$ be the set of all sums $a + b$ and let $T$ be the set of all products $ab$, where $a$ is a member of $A$ and $b$ is a member of $B$. Prove that if $p$ is the least upper bound of $A$ and if $q$ is the least upper bound of $B$, then $p + q$ is the least upper bound of $S$ and $pq$ is the least upper bound of $T$.

7. Show that there is no rational $r$ for which $r^2 = 5$.

8. Express as a single power of $x$:

$$\sqrt{x^3}, \ \sqrt{x^2}, \ \sqrt{x} \ \sqrt[3]{x}, \ (x^{1/2})^5, \ (x^{1/3} \cdot x^{1/4})^{1/5}, \ (x^2)^{3/5},$$

$$\frac{\sqrt[3]{x}}{\sqrt[6]{x}}, \ \frac{1}{x^7}, \ \frac{1}{x^{-7}}, \qquad \frac{x^{1/2}}{x^{1/3}} \cdot \frac{x^{-1/4}}{x^{1/6}} \ .$$

9. Give the base 2 logarithm of each of the following:

2, 1, 4, 8, 16, 32, 64, 1/2, 1/4, 1/8, $4^4$, $2 \cdot 8$, $32 \cdot 64$,

$(64)^{1/3}$, $(32 \cdot 64)/4^4$, $\sqrt[5]{64}$, $[(1/4) \cdot (1/8)]^{-5}$.

10. Show that if $n$ is any positive integer, then $2^n > n$. Is this true if $n$ is a positive rational?

# CHAPTER XII

## APPENDIX

### Introduction

IN THIS CHAPTER we re-examine from a slightly changed viewpoint some of the materials already treated. Instead of trying to detach intelligible portions from the whole body of mathematics, we shall try here to show how such portions fit into and suggest the larger science.

### Infinite Sets and the Cancellation Laws

In Chapter II we formulated the cancellation laws for addition and multiplication of cardinals but did not prove them. We discuss the proofs now.

Consider the set $A = \{1, 2, 3, \cdots\}$, $B = \{3, 4, 5, \cdots\}$. It is easy to see that $A$ can be put in one-to-one correspondence with $B$ by pairing 1 with 3, 2 with 4, 3 with 5, $\cdots$, $n$ with $n + 2$, etc. Thus, if $B$ had cardinal number $b$, $A$ would also also have cardinal number $b$. But since $A$ is the union $\{1, 2\} \cup B$, $A$ would also have cardinal number $2 + b$. It would follow that $2 + b = b$, and if the cancellation law for addition were valid we would have $2 = 0$. This, of course, would be a mathematical disaster. The cancellation law for multiplication has similar repercussions. Let $C$ be the set $\{2, 4, 6, \cdots\}$. Then $A$ and $C$ can be put in one-to-one correspondence by pairing 1 with 2, 2 with 4, 3 with 6, $\cdots$, $n$ with $2n$, etc. Thus, if $C$ had cardinal number $c$, $A$ would also have cardinal number $c$. But $A$ can be put in one-to-one correspondence with

$\{1, 2\} \times C$ and would therefore have cardinal number $2c$. It would follow that $2c = c$, and if the cancellation law for multiplication were valid, we would have $2 = 1$.

Since it is not the case that $2 = 0$ or that $2 = 1$, we must conclude either that the sets $A, B, C$ do not have cardinal numbers or that the cancellation laws are incorrect. Until about half a century ago, the first conclusion was the accepted one. The sets $A$, $B$, and $C$ were recognized as being infinite sets and were given no cardinal numbers. Nowadays, however, it is the cancellation laws that are abandoned, and the sets $A$, $B$, $C$, and indeed practically any set, can have a cardinal number. The cancellation laws do not hold in the arithmetic with infinite sets.

Actually, when we stated these laws originally we did so only for finite sets. There is nothing in the present discussion which makes dubious anything said about them in Chapter II. What the present discussion can accomplish is to clarify the task of proving the cancellation laws for finite sets. Knowing as we do now that these laws are not true for infinite sets, we see that no proof can be successful which does not exclude infinite sets. It turns out that the crucial aspect of a set in this connection is whether or not it can be put in one-to-one correspondence with one of its proper subsets. The sets we have called finite cannot be put in such one-to-one correspondence, while every set recognizable as being infinite can. Indeed, the formal definition used in present-day mathematics is that an infinite set is one that can be put in one-to-one correspondence with one of its proper subsets.

## Equal and Unequal Infinities

Early attempts to incorporate infinity into arithmetic took the form of denoting it by a symbol, usually "$\infty$", noting that the symbol introduced paradoxes into the usual laws of arithmetic, and then abandoning the whole venture. One of the strange features of present-day arithmetic with infinite sets is that there is not one but many infinities. Moreover, these infinities obey the five fundamental laws and in general have an orderly and interesting behavior. While we do not develop this subject further, we can prove that there are at least two distinct infinities.

DEFINITION: A set is *countable* if it can be put in one-to-one correspondence with the set of positive integers.

THEOREM: (a) *The set of integers is countable*; (b) *The set of rationals is countable*: (c) *The set of reals is not countable.*

*Proof*: To prove that the set of integers $\{0, \pm 1, \pm 2, \cdots\}$ is countable, we exhibit a one-to-one correspondence between that set and the set $\{1, 2, 3, \cdots\}$. One such correspondence pairs 1 with 0, 2 with 1, 3 with $-1, \cdots, 2n$ with $n$, $2n + 1$ with $-n$, etc.

To devise a one-to-one correspondence which shows that the set of rationals is countable, we represent the rationals in the form $\pm p/q$, where $p$ and $q$ are positive integers with no common factor. The following arrangement then furnishes a one-to-one correspondence:

| 1 | 2 | 3 | 4 | 5 | 6 | 7 | 8 | 9 | 10 | 11 | 12 | 13 | $\cdots$ |
|---|---|---|---|---|---|---|---|---|---|---|---|---|---|
| $\updownarrow$ | $\updownarrow$ | $\updownarrow$ | $\updownarrow$ | $\updownarrow$ | $\updownarrow$ | $\updownarrow$ | $\updownarrow$ | $\updownarrow$ | $\updownarrow$ | $\updownarrow$ | $\updownarrow$ | $\updownarrow$ | |
| $\frac{0}{1}$ | $\frac{1}{1}$ | $\frac{-1}{1}$ | $\frac{2}{1}$ | $\frac{1}{2}$ | $\frac{-1}{2}$ | $\frac{-2}{1}$ | $\frac{3}{1}$ | $\frac{1}{3}$ | $\frac{-1}{3}$ | $\frac{-3}{1}$ | $\frac{4}{1}$ | $\frac{3}{2}$ | $\cdots$ |

The scheme on which this is based is to write first all the different rationals $\pm p/q$ for which $p + q = 1$, then those for which $p + q = 2$, then those for which $p + q = 3$, etc. In this way, all the rationals are obtained and paired with a positive integer.

To prove that the reals are not countable, we must show that it is not possible to establish a one-to-one correspondence between the reals and the positive integers. We accomplish this by showing that in any correspondence alleged to be such a one-to-one correspondence, there is in fact at least one real that is not paired with a positive integer.

Let an alleged one-to-one correspondence pair

1 with $I_1 . a_{11} \, a_{12} \, a_{13} \cdots$
2 with $I_2 . a_{21} \, a_{22} \, a_{23} \cdots$
3 with $I_3 . a_{31} \, a_{32} \, a_{33} \cdots$
$\cdots\cdots\cdots\cdots\cdots\cdots\cdots$

What we are committed to doing in our proof is to scan this infinite list of reals and somehow produce a real absent from it. We offer the real $b = 0 . b_1 \, b_2 \, b_3 \cdots$ where

$b_1$ is 1 if $a_{11}$ is not 1, $b_1$ is 2 if $a_{11}$ is 1;
$b_2$ is 1 if $a_{22}$ is not 1, $b_2$ is 2 if $a_{22}$ is 1;
$b_3$ is 1 if $a_{33}$ is not 1, $b_3$ is 2 if $a_{33}$ is 1;

. . . . . . . . . . . . . . . . . . . . . . . . . . . . . . . . .

We have certainly defined a real number, since we have specified a suitable entry for each decimal place. That it is not on the list follows readily from the construction. If it were on the list, it would be the real paired with some integer $n$, namely

$$I_n . a_{n1} \, a_{n2} \cdots .$$

Clearly this is impossible, because the entry $b_n$ in $b$ is by construction different from the entry $a_{nn}$ in the listed real.

## Peano Axioms for the Natural Numbers

The approach to the familiar symbols $1, 2, 3, \cdots$ set forth in Chapter II is a far more ambitious one than that of many mathematicians. Our attempt there was to construct objects that could be recognized as having all the properties these symbols suggest. Mathematicians are often willing to bypass the construction and start with the numbers themselves. However, they do not simply carry them over without change from everyday language into the formal language, because many of the properties of everyday language are fuzzy, and unsuitable for mathematical discussion. What they do is to announce that they are going to consider a certain set which is to have certain stated properties. The stated properties are chosen in such a way as to guarantee that the remaining desired properties follow from them as logical consequences. The stated properties are sometimes called *axioms,* and sometimes called *assumptions.*

A famous treatment of the symbols $1, 2, 3, \cdots$ from this point of view is due to the mathematician Peano. We give his axioms for the *natural numbers.* His axioms refer to an otherwise unspecified set $S$. They are:

I.  1 is a member of $S$.

II.  For each member $a$ of $S$ there is a certain member $a^+$ of $S$ called the *successor of a.*

III. For each member $a$ of $S$, $a^+$ is not 1.

IV. If $a$ and $b$ are members of $S$, and if $a^+ = b^+$ then $a = b$.

V. Each subset $T$ of $S$ that meets the following conditions is $S$ itself:

(a) 1 is a member of $T$.

(b) There is no $a$ for which $a$ is a member of $T$ and $a^+$ is not.

It is possible to define addition and multiplication for members of the set $S$ in such a way that the resulting number system is isomorphic to the system of positive integers. Addition can be defined by the rules

$$x + 1 = x^+$$
$$x + y^+ = (x + y)^+$$

and multiplication by the rules

$$x \cdot 1 = x$$
$$x \cdot y^+ = x \cdot y + x,$$

where $x$ and $y$ are variables with domain $S$.

We do not show in detail that the endowment of the set $S$ with these operations produces the familiar number system. We merely state that it can be done, neatly and satisfactorily. The only possible drawback to the procedure is that it has a hypothetical character. There is no evidence offered to prove that there is such a set $S$; evidence is offered exclusively to show that if there is such a set it has certain properties. Whether or not this hypothetical feature is to be deplored, it is encountered throughout the body of mathematics.

## Proof by Mathematical Induction

In Chapter II we faced the problem of proving that each statement of an infinite list of statements is valid. Mathematical induction is the name given to a particularly efficient way of doing this. To use the method, one must have an infinite list of statements that have been put in one-to-one correspondence with the natural numbers. Let us denote the statements by $S_1, S_2, S_3, \cdots$. The object of the method is to prove that each of these statements is valid. Two simple steps constitute the method.

1. Show that $S_1$ is valid.

2. Show that for each $n$ for which $S_n$ is valid, $S_{n+1}$ is also valid. Before we examine the justification for the method, let us use it in an example.

Let us take for $S_1, S_2, \cdots$ the following:

$S_1: 1 = 1^2,$
$S_2: 1 + 3 = 2^2,$
$S_3: 1 + 3 + 5 = 3^2,$
. . . . . . . . . . . . . . . . .
$S_n: 1 + 3 + 5 + \cdots + 2n - 1 = n^2,$
. . . . . . . . . . . . . . . . . . . . . . . . . . . . .

In other words, we are going to try to show that if we add up any given number of consecutive odd numbers beginning with 1, the sum is the square of that given number. We must show first that $S_1$ is valid, that is, that $1 = 1^2$. We do this by referring to the multiplication table. The second step is to show that if $S_n$ is valid, so is $S_{n+1}$; that is, if

$$1 + 3 + 5 + \cdots + 2n - 1 = n^2$$

then

$$1 + 3 + 5 + \cdots + 2n - 1 + 2n + 1 = (n + 1)^2.$$

We have

$$1 + 3 + 5 + \cdots + 2n - 1 + 2n + 1$$
$$= (1 + 3 + 5 + \cdots + 2n - 1) + 2n + 1 \quad \text{(GA)}$$
$$= n^2 + 2n + 1 \quad \quad \text{($S_n$ assumed valid)}$$
$$= (n + 1)^2. \quad \quad \quad \quad \quad \quad \text{(GD)}$$

This is the complete proof.

Let us now examine the justification of the method. It is based on the property of the natural numbers contained in Axiom 5 above. To prove that each of $S_1, S_2, \cdots$ is valid is clearly equivalent to showing that the set of natural numbers $r$ for which $S_r$ is valid is the entire set of natural numbers. This is what the method of mathematical induction does. To see this, define a subset of the natural numbers as consisting of those numbers $r$ for which $S_r$ is valid. Step 1 of mathematical induction tries to show that the number one is in this subset, and step 2 tries to show that the remaining conditions of Axiom 5 are applicable. If these steps can be carried out, then the subset is the entire set, and all the statements are valid.

## Axioms, Groups, Rings, and Fields

Peano's axioms have an appearance of plausibility which is in accord with the everyday use of the word axiom. Something axiomatic is, in everyday usage, something self-evident. A beginner reading Peano's axioms and deductions based on them might feel that the mathematicians are having some ponderous joke in attempting to prove what is known very well at the outset. In so surmising he would miss the whole point. Actually the question is not whether natural numbers have certain properties, but whether certain otherwise totally unspecified objects, known only to have the number properties stated in the axioms, must necessarily possess all the other properties as well. The mathematician is not pretending ignorance about what is, in fact, familiar but is rather obtaining information about what is, in fact, unfamiliar. The procedure contributes greatly to the success of mathematics in its applications. We saw a glimpse of it in our proof of the three laws $GA, GM, GD$ from the five basic laws. It was not particularly striking that the three laws held for cardinals. What was striking was that they could be shown to hold in any system whatsoever in which the five laws were valid. It is a standard mathematical practice to say at the outset as little as possible about the objects to be studied. This is not prompted by timidity or stinginess. The fewer the limitations one imposes on the objects of one's study, the greater is the applicability of the results one obtains.

We now describe some objects which have been studied extensively from the axiomatic point of view, and which are closely related to objects studied in this book. They are *groups*, *rings*, and *fields*.

DEFINITION: A *group* is a set, with an operation $\odot$ that is subject to the following axioms:

  I. The operation $\odot$ is associative.

  II. The operation $\odot$ has an identity element.

  III. Each member of $S$ has an inverse element.

Those groups in which the operation is commutative are known as commutative groups.

DEFINITION: A *ring* is a set $S$ that has two operations $\oplus$ and $\odot$ subject to the following axioms:

  I. $S$ is a commutative group relative to $\oplus$.

  II. $\odot$ is an associative operation.

  III. $\odot$ is distributive relative to $\oplus$.

If, in a ring, the multiplication operation is commutative and has an identity element, and if the cancellation law for multiplication holds, then the ring is said to be an *integral domain*.

DEFINITION: A *field* is a set $S$ which has two operations $\oplus$ and $\odot$ subject to the following axioms:

  I. $S$ is a commutative group relative to $\oplus$.

  II. $S$, with the identity element relative to $\oplus$ deleted, is a commutative group relative to $\odot$.

  III. $\odot$ is distributive relative to $\oplus$.

*Examples*: The integers, the rationals, and the reals are groups relative to their addition operations. The integers are an integral domain, and the rationals and the reals are each fields.

The reader thus has some experience with individual groups, rings, and fields. He can certainly understand the meaning of the questions that follow. They are all constructed by analogy with known facts and are offered as legitimate and typical mathematical questions.

Can the construction which gave the rationals from the integers be applied to any integral domain? (The answer is *yes*.)

Must the construction yield a field? (The answer is *yes*.)

Does the new field contain a subset isomorphic to the original integral domain? (The answer is *yes*.)

Can the construction that yielded the reals from the rationals be applied to any field? (The answer is *no*.)

We also raised and answered similar questions in our chapter on equations and in our chapter on order.

We have thus far given the answers to our questions. This may give the reader the totally false impression that mathematics has answers to all its questions. This is far from the case. The material of this book is a tiny portion of the mathematics that is known, and this again is a tiny portion of what lies ahead to be discovered.

## Exercises

1. Each of the following suggests an infinite list of statements. Formulate them and prove them by mathematical induction.

   a) $2 = 2 \cdot 1^2$
   $2 + 6 = 2 \cdot 2^2$
   $2 + 6 + 10 = 2 \cdot 3^2$
   $2 + 6 + 10 + 14 = 2 \cdot 4^2$.

   b) $ax = ax$
   $(a + b)x = ax + bx$
   $(a + b + c)x = ax + bx + cx$
   $(a + b + c + d)x = ax + bx + cx + dx$.

   c) $x^n \cdot x = x^{n+1}$
   $x^n \cdot x^2 = x^{n+2}$
   $x^n \cdot x^3 = x^{n+3}$
   $x^n \cdot x^4 = x^{n+4}$.

   d) $1/(1 \cdot 2) = 1/2$
   $1/(1 \cdot 2) + 1/(2 \cdot 3) = 2/3$
   $1/(1 \cdot 2) + 1/(2 \cdot 3) + 1/(3 \cdot 4) = 3/4$
   $1/(1 \cdot 2) + 1/(2 \cdot 3) + 1/(3 \cdot 4) + 1/(4 \cdot 5) = 4/5$.

2. Let $n$ be a positive integer, and let $S$ be the set

$$\{0, 1, \cdots, n - 1\}.$$

Let $\oplus$ be the operation on $S$ defined by assigning to each pair of members of $S$ what is obtained from them by first adding them in the usual way, and then dividing by $n$ and retaining the remainder.

a) Show that $S$ is a group relative to $\oplus$.

b) Define an operation $\odot$ on $S$ by imitating $\oplus$ with multiplication.

c) Show that $S$ is a commutative ring relative to $\oplus$ and $\odot$.

d) Show that if $n$ is a prime, then $S$ is a field. (NOTE: The reader who doesn't know what a prime is or whose acquaintance with primes is meager, is advised to omit this problem).

3. Show that

a) the cardinals do not form a group relative to $+$.

b) the cardinals do not form a group relative to $\cdot$

c) the integers do form a group relative to $+$.

d) the integers do not form a group relative to $\cdot$

e) the rationals do form a group relative to $+$.

f) the rationals do not form a group relative to $\cdot$

g) the rationals, with zero omitted, do form a group relative to $\cdot$

# INDEX

# INDEX